OF GOD AND DICE

QUOTES FROM EMINENT SCIENTISTS SUPPORTING A CREATOR

GORDON LEIDNER

Of God and Dice

Copyright © 2012 by Gordon Leidner

ISBN: 1470041502
ISBN 13: 9781470041502

TO JEAN, WITH LOVE.

TABLE OF CONTENTS

INTRODUCTION

"Why is there *something* rather than *nothing?*"[1]

This question, attributed to seventeenth century German mathematician Gottfried Leibniz, has challenged scientists and philosophers for centuries. To Leibniz the answer was obvious—God created the universe *ex nihilo*. But to the atheist, the answer is both complicated and elusive.[2]

This book will show how skeptics of the existence of God sometimes develop extremely complicated cosmological theories—a number of them rather bizarre—in their attempts to obviate the need for a supernatural Creator. To make the math in their theories work, they frequently begin with unconventional assumptions and significantly "fine tune" their boundary conditions. Even though the results cannot be substantiated by observation and scientific experiment, conclusions are drawn and works are published. After this, the "new atheists" take over, choosing the findings that support their point of view and rushing their own books or articles into print. Then they state, once again, that modern science has *proven* there was no Creator.

The atheists' claim that believing in an "un-provable" God requires blind faith, whereas believing their conclusions is to accept proven, scientific fact. *Of God and Dice* will show that these "facts" may themselves be based on faith, and are sometimes quite chimerical.

The purpose of this book is to demonstrate that there are many respected scientists who do *not* believe science has "disproven" the existence of God. They believe, in fact, the opposite: God exists, and is the Creator of the universe. This is accomplished by using their own words, organized under topics considered appropriate for the task. Many of the quotations have the purpose of illustrating the scientists' support of the existence of God, while others have more indirect purposes—such as illustrating the fallacies of theories atheists use in their arguments against God.

I have limited the quotations to those I believe to be from "mainstream" scientists. These include well known giants of history such as Albert Einstein, Isaac Newton, Max Planck, and Johannes Kepler. They also include many lesser known, but still highly influential scientists such as Werner Heisenberg, Arno Penzias, Charles H. Townes, and Francis Collins. Many are Nobel laureates. All are academically qualified and well respected in their fields of study.

Of God and Dice does not presume to lay the science vs. religion argument to rest. It does not attempt to provide a balanced, two-sided argument. The book neither supports nor opposes evolution or creation science. *Of God and Dice* focuses on the readers in the middle—those who question the existence of God.

Since Albert Einstein's allegiance is claimed by the atheists and faithful alike, a note about his religious beliefs is in order. It is a well-known fact that Einstein did not believe in a "personal" God, i.e., one that concerns Himself with the affairs of mankind or answers prayer. But it is equally well-known that he vehemently denied that he was an atheist. The quotations from Einstein show that he was very supportive of the concept of a creator God. This distinction could also be made regarding some of the others quoted herein, but I do not attempt to elaborate on the religious beliefs of the scientists. Suffice it to say that many are Christian, some are Jewish, a few follow other major world religions, some are theists, and some are simply deists.

The title of the book, *Of God and Dice*, is derived from a famous quote by Albert Einstein. Einstein believed that God created a universe of perfect, understandable order. He believed that God meant for His creation to be completely understood by man, using tools such as mathematics and observation. In fact, scientists thought this is the way things were—until quantum theory (also called quantum mechanics) was introduced by Max Planck in the early twentieth century. Planck and others proved that the universe, at the subatomic (quantum) level, was not completely predictable. Einstein's statement "God does not throw dice" reflected his frustration over the new theory that introduced unexplainable mystery into the behavior of subatomic particles, and hence, the universe.

Of God and Dice is organized as follows: at the beginning of each chapter, I provide a brief introduction that offers some context for the scientists' quotes that follow. Sources for the quotes are in end notes in the back of the book, and short biographical material for each scientist is in Chapter Ten.

Finally, I should present my personal definition of God. I consider God to be *the omniscient, omnipotent, omnibenevolent, supernatural creator and sustainer of both the physical universe and the laws that govern it. He is the Supreme Being who, because of His inherent nature, is eternal—needing no prior cause for His existence.*

Gordon Leidner
April, 2012
www.OfGodandDice.com

ONE

THE LIMITATIONS OF SCIENCE

Science can only ascertain what is, but not what should be.

— *Albert Einstein*[3]

Contrary to what many people think, science cannot answer every question. This is primarily due to its inherent limitations in process. According to the Oxford English Dictionary, the scientific method is "a method of procedure that has characterized natural science since the 17th century, consisting of systematic observation, measurement and experiment, and the formulation, testing and modification of hypotheses."

Scientific researchers start with a question, make observations, form a hypothesis, test the hypothesis by experiment, analyze the data, draw conclusions, and if necessary retest. If this full cycle is not fulfilled, it's technically not "science." True science requires observation and repeatable, verifiable experiments.

The existence of God, therefore, cannot be refuted by science, because there are no experiments that can be performed for proof or disproof. Likewise, many of the currently popular scientific theories, such as "multiple universes," cannot be proven either. As will be seen in later chapters, no scientific experiment can be done to verify the existence of universes other than our own because there is no way of observing them. Atheists, who scoff at believers in God as people dependent on "faith," actually have a great deal in common with those they scorn. Atheistic scientists are using unprovable theories to support their own beliefs. As will be shown, both positions require "faith."

The following quotes emphasize science's inability to disprove the existence of a creator God.[4]

***

Science can tell you what the laws of physics are, but it cannot tell you why they exist. Science cannot tell you why the universe exists. And above all it cannot tell you whether or not God exists. ...These limitations cannot be changed by future advances in science; they are fundamental in its nature. So we can expect many advances in science in the future.. but we cannot expect it to solve ethical or moral or metaphysical issues. *George Ellis*[5]

***

But what of the concept of "proof"? Certainly, many would argue that proofs give scientific ideas a kind of absolutism and universalism that religion lacks. In truth, however, we can never prove anything completely. Even scientifically and mathematically, we can never be absolutely sure our conclusions are correct. Our science is based on postulates, or assumptions, which, like faith, we may believe in firmly, but cannot prove absolutely. *Charles H. Townes*[6]

***

Science doesn't solve life's problems. It may provide description but it doesn't provide meaning. Science may provide some clues to meaning, but very weak ones. This is one of the reasons I specifically say we cannot know the mind of God. *Arno Penzias*[7]

***

Science without religion is lame, religion without science is blind. *Albert Einstein*[8]

***

Science cannot solve the ultimate mystery of nature. And that is because, in the last analysis, we ourselves are part of nature and therefore part of the mystery that we are trying to solve. *Max Planck*[9]

***

Why are we here? Does the universe have a purpose? Whence comes our knowledge of good and evil? These mysteries, and a hundred others like them, are beyond the reach of science. They lie on the other side of the border, within the jurisdiction of religion. *Freeman Dyson*[10]

***

Whence come I and whither go I? That is the .. unfathomable question, the same for every one of us. Science has no answer to it. *Erwin Schrödinger*[11]

***

Science is good at telling us how, but not so good on the why. *Paul Davies*[12]

***

We cannot discover, through science alone .. the answers to the questions "Why is there life anyway?" and "Why am I here?" *Francis Collins*[13]

***

Science can never answer the religious questions. You still have 'what came before?' and you can ask 'who designed it all?' *George Smoot*[14]

***

I acknowledge freely that religion goes beyond science in that it grapples with timeless questions that science cannot answer. *John A. Russell*[15]

***

While science itself cannot address or prove the existence or non-existence of God, there are other compelling reasons, looking at nature and experience as a whole, for many people to believe in God. And from that perspective of faith, the Creation itself will reflect the nature of God. *Jennifer Wiseman*[16]

***

Religions and science answer different questions about the world. Whether there is a purpose to the universe or a purpose for human existence are not questions for science....Science is a way of knowing about the natural world. It is limited to explaining the natural world through natural causes. Science can say nothing about the supernatural. Whether God exists or not is a question about which science is neutral. *National Academy of Sciences*[17]

***

Science cannot either prove or disprove religion. Religion is founded on faith. It seems to me that when confronted with the marvels of life and the universe, one must ask why and not just how. The only possible answers are religious. For me that means Protestant Christianity, to which I was introduced as a child and which has withstood the tests of a lifetime. *Arthur Shawlow*[18]

***

One of the most important questions in the history of human thought, namely, 'Why do we exist? or, how did we get here? turns out to have an answer that is beyond the reach of scientific inquiry. *Robert Jastrow*[19]

***

Science needs religion in order to account for its origin and its successes. I discussed this view with Einstein when I did research at the Institute of Advanced Study at Princeton in 1932 and remember his comment: "The discovery of a fundamental, verified law of nature is an inspiration of God." *Henry Margenau*[20]

***

Science is powerless to answer questions such as "Why did the universe come into being?" *Francis Collins*[21]

***

The scientific method itself has fundamental limits, and many important areas lie outside those limits. *George Ellis*[22]

***

Go back further still, beyond the moment of creation, what then? What was there before the big bang? What was there before time began? Facing this, the ultimate question, challenges our faith in the power of science to find explanations of nature. The existence of a singularity—in this case the given, unique state from which the universe emerged—is anathema to science because it is beyond explanation. There can be no answer to why such a state existed. Is this, then, where scientific explanation breaks down and God takes over, the artificer of that singularity, that initial simplicity? *George Smoot*[23]

***

It is only in dealing with the ultimate questions that science and logic may fail us. I am not saying that science and logic are likely to provide the wrong answers, but they may be incapable of addressing the sort of "why" (as opposed to "how") questions we want to ask. *Paul Davies*[24]

***

Science by itself does not either confirm or deny a spiritual or religious understanding of the universe. Those making that link are following an illegitimate line of argument that cannot survive critical scrutiny. *George Ellis*[25]

***

The scientific picture of the real world around me is very deficient. It gives a lot of factual information, puts all our experience in a magnificently consistent order, but it is ghastly silent about all and sundry that is really near to our heart, that really matters to us. It cannot tell us a word about red and blue, bitter and sweet, physical pain and physical delight; it knows nothing of beautiful and ugly, good or bad, God and eternity. Science sometimes pretends to answer questions in these domains, but the answers are very often so silly that we are not inclined to take them seriously. *Erwin Schrödinger*[26]

***

We can never understand what events are, but must limit ourselves to describing the patterns of events in mathematical terms; no other aim is possible. Physicists who are trying to understand nature may work in many different fields and by many different methods; one may dig, one may sow, one may reap. But the final harvest will always be a sheaf of mathematical formulae. These will never describe nature itself. ... [Thus] our studies can never put us into contact with reality. *Sir James Jeans*[27]

***

Now we see how the astronomical evidence supports the biblical view of the origin of the world....the essential elements in the astronomical and biblical accounts of Genesis are the same. Consider the enormousness of the problem: Science has proved that the universe exploded into being at a certain moment. It asks: 'What cause produced this effect? Who or what put the matter or energy into the universe?'.. And science cannot answer these questions. *Robert Jastrow*[28]

TWO

THE BIBLE'S IMPACT

> The highest principles for our aspirations and judgments are given to us in the Jewish-Christian religious tradition.
>
> *— Albert Einstein*[29]

Before the eighteenth century, most of Western Civilization's best known scientists were firm believers in the God of the Bible. These included giants of astronomy and physics such as Copernicus, Galileo, Kepler, and Newton. It also included famous scientists of other scientific disciplines, such as Bacon and Descartes.

What was the impact of the Bible on scientific progress in Western Civilization? Many studies have been done on this subject, and it has been debated from various viewpoints. Most conclude that the Bible was a significant influence in that it: 1) stated that there had been a beginning to the universe, caused by an omnipotent, benevolent creator; 2) this creation was done in an orderly, timely manner and proclaimed to be "good;" 3) there was an ordained timeline, with an eventual end to the earth as we know it; 4) man was commanded to "subdue" the earth; 5) man's chief end was to glorify God and enjoy Him forever. Since man was to glorify the supernatural Creator while subduing the good creation, the study of science became an effective means of accomplishing both tasks.

Another viewpoint regarding the impact of the Bible on science stems from a comparison of the scientific advances between pre-eighteenth century Western and Eastern societies. Some have concluded that the philosophical position of Judeo-Christian society was much more conducive to scientific exploration.[30] Hence, most of the significant pre-eighteenth century advances in physics and astronomy were made in the Christian-dominated west.

***

If God does not exist, one will lose nothing by believing in him, while if he does exist, one will lose everything by not believing. *Blaise Pascal.*[31]

***

There is nothing I can find out and long to know with greater urgency than this. Can I find God, whom I can almost grasp with my own hands in looking at the universe, also in myself? *Johannes Kepler*[32]

***

To know the mighty works of God, to comprehend His wisdom and majesty and power; to appreciate, in degree, the wonderful workings of His laws, surely all this must be a pleasing and acceptable mode of worship to the Most High, to whom ignorance cannot be more grateful than knowledge. *Nicolaus Copernicus*[33]

***

I consider it a favor of heaven that since childhood a faith is planted deep in my innermost being, a faith in the Almighty and All-good not to be shattered by anything. *Max Planck*[34]

***

All human discoveries seem to be made only for the purpose of confirming more strongly the truths come from on high, and contained in the sacred writings. *Sir John Frederick William Herschel*[35]

***

But the context of religion is a great background for doing science. In the words of Psalm 19, 'The heavens declare the glory of God and the firmament showeth his handiwork." Thus scientific research is a worshipful act, in that it reveals more of the wonders of God's creation. *Arthur Shawlow*[36]

***

As a religious person, I strongly sense the presence and actions of a creative Being far beyond myself and yet always personal and close by. *Charles H. Townes*[37]

***

My lack of doubt about God's existence is at a personal level, in that I am conscious of his presence every moment of the day. I could no more doubt his existence than doubt my own. *Ken Freeman*[38]

***

He hath set his testimony in the heavens. *Michael Faraday*[39]

***

Science forms a valuable part of human life, but it is not the basis for a whole human life. We shall always need to study and teach ethics, aesthetics, and philosophy, as well as science—and this should include comparative religion if you want a whole human being. Those who claim science will supplant any or all of them are indulging in a little fantasy. Be kind to them, but don't take them seriously. *George Ellis*[40]

***

The Supreme God is ... eternal and infinite, omnipotent and omniscient; that is, his duration reaches from eternity to eternity; his presence from infinity to infinity; he governs all things, and knows all things that are or can be done.... We know him only by his most wise and excellent contrivances of things, and final causes; we admire him for his perfections; but we reverence and adore him on account of his dominion. *Sir Isaac Newton*[41]

***

The undevout astronomer must be mad. *Sir William Herschel*[42]

***

Certainly, the idea of God, or a supremely perfect being, is one that I find within me just as surely as the idea of any shape or number. *Rene Descartes*[43]

***

I find a need for God in the universe and in my own life. *Arthur Schawlow*[44]

***

[God] is the kind Creator who brought forth nature out of nothing. *Johannes Kepler* [45]

***

We ought to value the privilege of knowing God's truth far beyond anything we can have in this world. *Michael Faraday*[46]

***

I believe, with the Westminster Divines and their pre-decessors ad Infinitum that "Man's chief end is to glorify God and to enjoy him forever." *James Clerk Maxwell*[47]

***

The last service that, I hope…is to induce men to pay their admiration, their praises, and their thanks, directly to God himself; who is the true and only creator of the sun, moon, earth, and those other creatures, that men call the works of nature. *Sir Robert Boyle*[48]

***

Johannes Kepler

Blaise Pascal

Renee Descartes

Sir Isaac Newton

Many scientists are also people with quite conventional religious faith, I, a physicist, am one example. I believe in God as both creator and friend. That is, I believe that God is personal and interacts with us. *William Phillips*[49]

***

The infinite abyss can only be filled with something that is infinite and unchanging—in other words, by God himself. God alone is our true good. *Blaise Pascal*[50]

***

By investigating God's majestic and awesome creation, science can actually be a means of worship. *Francis Collins*[51]

***

I invite you to examine the snapshot provided by half a century's worth of astrophysical data and see what the pieces of the universe actually look like... The best data we have are exactly what I would have predicted had I nothing to go on but the five books of Moses, the Psalms, the Bible as a whole. *Arno Penzias*[52]

***

God has, in fact, written two books, not just one. Of course, we are all familiar with the first book he wrote, namely Scripture. But he has written a second book called creation. *Francis Bacon*[53]

***

For myself, faith begins with the realization that a supreme intelligence brought the universe into being and created man. It is not difficult for me to have this faith, for it is incontrovertible that where there is a plan there is intelligence. An orderly, unfolding universe testifies to the truth of the most majestic statement ever uttered: "in the beginning God…" *Arthur Compton*[54]

***

God is a concept, which I need to cohere my total experience. Christianity comes nearest to the formal expression of this for me. You've got to have something other than just scientific laws. More science is not going to answer all the questions that we ask. *Antony Hewish*[55]

***

Posterity will one day laugh at the foolishness of modern materialistic philosophers. The more I study nature, the more I stand amazed at the work of the Creator. I pray while I am engaged at my work in the laboratory. *Louis Pasteur*[56]

***

I worship an unseen God…getting to know him…changed my whole world-view. *Sir Robert Boyd*[57]

***

I have looked into most philosophical systems and I have seen that none will work without a God. *James Clerk Maxwell*[58]

***

When you're doing physics, you're wrestling with a champ. You're trying to find out how God made the world, just like Jacob wrestling with the angel. *Isidor Isaac Rabi*[59]

***

I am quite beside myself with wonder and infinitely grateful to God that it has pleased Him to permit me to discover such great marvels. *Sir James Jeans*[60]

***

So vast, without any question, is the Divine Handiwork of the Almighty Creator! *Nicolaus Copernicus*[61]

***

I thank thee, Lord God our Creator, that thou allowest me to see the beauty in thy work of creation. *Johannes Kepler*[62]

***

Max Planck, if I understand him rightly, has used [his] freedom and come down squarely on the side of the Christian tradition. His thoughts and actions, particularly as they affect his personal relationships, fit perfectly into the framework of this tradition, and no one will respect him the less for it. *Werner Heisenberg*[63]

THREE

SCIENCE AND RELIGION: HEAVENLY MARRIAGE?

> I found it difficult to imagine there could be real conflict between scientific truth and spiritual truth. Truth is truth.
>
> — *Francis Collins*[64]

A shift occurred in western scientists' attitudes toward God after the Age of Enlightenment in the eighteenth century. Belief in the success of their methodology, based upon Bacon's Scientific Method, became to them a sufficient motivator for study. After the development of classical physics with Isaac Newton, science was increasingly seen as "deterministic." Scientists believed they would soon be able to understand and predict *everything,* which led to the gradual separation of science and religion.

During this time, scientists saw the universe as static and predictable, and interest in the biblical account of creation diminished. But as will be shown in subsequent chapters, several scientific developments in the twentieth century changed everything, causing a renewed interest in God by many leading scientists.

Although fewer scientists today would say that the unraveling of the mysteries of God's creation is their primary motivator, many see value in the simultaneous pursuit of science and religion. A common perception of the modern scientist is that religion asks the "why" questions, such as "Why are we here?" and science pursues the "how" questions, such as "How did the universe evolve?" Noteworthy is the fact that science in the twentieth and twenty-first centuries has been significantly impacted by the discoveries of Christian scientists such as Max Planck, Arthur H. Compton, Charles H. Townes, Arthur Schawlow, and Francis Collins.

***

It was not by any accident that the greatest thinkers of all ages were deeply religious souls, even though they made no public show of their religious feelings. *Max Planck*[65]

***

In the view of such harmony in the cosmos which I, with my limited human mind, am able to recognize, there are yet people who say there is no God. But what makes me really angry is that they quote me for support for such views. *Albert Einstein*[66]

***

The first gulp from the glass of natural sciences will turn you into an atheist, but at the bottom of the glass God is waiting for you. *Werner Heisenberg*[67]

***

I find that religion and science present no conflict. Both proceed from acts of faith. ..I firmly agree that current questions now arising in cosmology, elementary particle physics, and microbiology have an obvious metaphysical or religious content. *Robert A. Naumann*[68]

***

I see no conflict between science and religion; they are basically addressing different questions. Science asks how, and religion asks why. *John G. Phillips*[69]

***

Science wants to know the mechanism of the universe, religion–the meaning. The two cannot be separated. *Charles H. Townes*[70]

***

This much I can say with definiteness – namely, that there is no scientific basis for the denial of religion – nor is there in my judgment any excuse for a conflict between science and religion, for their fields are entirely different. Men who know very little of science and men who know very little of religion do indeed get to quarreling, and the onlookers imagine that there is a conflict between science and religion, whereas the conflict is only between two different species of ignorance. *Robert Millikan*[71]

***

I .. do not believe that there are any collisions between what I believe as a Christian, and what I know .. as a scientist. *Francis Collins*[72]

***

Both Religion and science require a belief in God. For believers, God is in the beginning, and for physicists He is at the end of all considerations… To the former He is the foundation, to the latter, the crown of the edifice of every generalized world view. *Max Planck*[73]

***

I think that religion and science are two different approaches to understanding existence. Their domains do not overlap completely. *Vera Kistiakowsky*[74]

***

Science, with its experiments and logic, tries to understand the order or structure of the universe. Religion, with its theological inspiration and reflection, tries to understand the purpose or meaning of the universe. These two are cross-related. Purpose implies structure, and structure ought to be interpretable in terms of purpose. At least this is the way I see it. I am a physicist. I also consider myself a Christian. *Charles H. Townes*[75]

***

Science and theology are both concerned with the search for truth. In consequence, they complement each other rather than contrast one another. Of course, the two disciplines focus on different dimensions of truth, but they share a common conviction that there is truth to be sought. *John Polkinghorne*[76]

***

I have never found it possible to dismiss the content of religious thinking as simply part of an outmoded phase in the consciousness of mankind, a part we shall have to give up from now on. Thus in the course of my life I have been repeatedly compelled to ponder on the relationship of these two regions (religion and science) of thought, for I have never been able to doubt the reality of that to which they point. *Werner Heisenberg* [77]

***

Science brings men nearer to God. *Louis Pasteur*[78]

***

It is true, that a little philosophy [science] inclineth man's mind to atheism; but depth in philosophy bringeth men's minds about to religion. *Francis Bacon* [79]

***

I have been both religious and scientifically oriented since childhood, and the two realms have always fit together from my point of view. *Charles H. Townes*[80]

***

Do not be afraid of being free thinkers. If you think strongly enough you will be forced by science to the belief in God, which is the foundation of all religion. You will find science not antagonistic but helpful to religion. *Lord Kelvin*[81]

***

Science is possible because the universe is a divine creation. *John Polkinghorne*[82]

***

There can never be any real opposition between religion and science; for the one is the complement of the other. Every serious and reflective person realizes, I think, that the religious element in his nature must be recognized and cultivated if all the powers of the human soul are to act together in perfect balance and harmony. *Max Planck*[83]

***

Human well-being and all human progress rest at bottom upon two pillars, the collapse of either one of which will bring down the whole structure. These two pillars are the cultivation and the dissemination throughout mankind of 1) the spirit of religion, and 2) the spirit of science (or knowledge). *Robert Millikan*[84]

***

The popular perception is that science had proven the Bible wrong. In reality, the claim of Earth's centrality had nothing to do with the Bible. *{Regarding the Catholic Church's inquisition of Galileo} Gerald Schroeder*[85]

***

Sometimes people ask if religion and science are not opposed to one another. They are; in the sense that the thumb and fingers of my hand are opposed to one other. It is an opposition by means of which anything can be grasped. *William Bragg*[86]

***

Arthur H. Compton

***

In their essence there can be no conflict between science and religion. Science is a reliable method of finding truth. Religion is the search for a satisfying basis for life. *Arthur H. Compton*[87]

***

The impossibility of real science and real religion ever conflicting becomes evident when one examines the purpose of science and the purpose of religion. The purpose of science is to develop – without prejudice or preconception of any kind – knowledge of the facts, the laws and the processes of nature. The even more important task of religion, on the other hand, is to develop the consciences, the ideals, and the aspirations of mankind. *Robert Millikan*[88]

***

I believe that science and religion deal with different kinds of questions using different (but not completely different) methods. Science addresses questions about how things work, the history of development of the universe, and the like. Religion addresses questions about ultimate meaning, about what ought to be the relationship among people, and between people and God. .. I also believe that God is revealed in part through the observations we make on the creation. *William Phillips*[89]

***

A scientific discovery is also a religious discovery. There is no conflict between science and religion. Our knowledge of God is made larger by every discovery we make about the world. *Joseph H. Taylor*[90]

***

Science can have a purifying effect on religion, freeing it from beliefs from a pre-scientific age and helping us to a truer conception of God. At the same time, I am far from believing that science will ever give us the answers to all our questions. *Nevill Mott*[91]

***

I must confess that I myself do not feel altogether happy about this separation [science and religion]. I doubt whether human societies can live with so sharp a distinction between knowledge and faith. *Werner Heisenberg*[92]

***

I believe that the more thoroughly science is studied, the further does it take us from anything comparable to atheism. *Lord Kelvin*[93]

***

I believe in Spinoza's God who reveals himself in the orderly harmony of what exists, not in a God who concerns himself with fates and actions of human beings. *Albert Einstein*[94]

***

Spinoza was no atheist; he believed he had a logical proof that God must exist. Because he identified God with the physical universe, this amounted to a proof that our particular universe must also exist. *Paul Davies*[95]

***

We are trying to discover what the Lord thought; of course we miserably fail most of the time, but sometimes there is a great satisfaction in seeing a little bit of the truth. *Abdus Salam*[96]

***

Physics filled me with awe, put me in touch with a sense of original causes. Physics brought me closer to God. *Isidor Isaac Rabi*[97]

***

In my view, much of the [world's] trouble lies in the philosophical bases underpinning modern science. Originally, science arose in a religious context. Its pioneers saw nature as the handiwork of an All-knowing Almighty God. They also saw humankind's role to be the stewardship of the Earth. But over the past two centuries this picture changed. Science in its modern form is silent about God and about our responsibilities toward ourselves, society, and the world. The prevalent view now concentrates on exploring nature and putting it at the service of human beings. *Mehdi Golshani*[98]

***

What really interests me is whether God had any choice in the creation of the world. *Albert Einstein*"[99]

***

At this moment it seems as though science will never be able to lift the curtain on the mystery of creation. For the scientist who has lived by his faith in the power of reason, the story ends like a bad dream. He has scaled the mountain of ignorance; he is about to conquer the highest peak; as he pulls himself over the final rock, he is greeted by a band of theologians who have been sitting there for centuries. *Robert Jastrow*[100]

FOUR

QUANTUM THEORY AND SUPERNATURAL FORCE

In seeking wisdom thou art wise; in imagining that thou hast attained it - thou art a fool.

— Lord Chesterfield[101]

The deterministic methods of Newtonian physics were made obsolete in the early twentieth century by the new "quantum theory."[102] Max Planck and Albert Einstein demonstrated that light, at that time understood as an electromagnetic wave, also behaved as if it were made up of discreet particles called photons. Erwin Schrödinger and Werner Heisenberg subsequently developed theories which further illustrated that subatomic entities such as electrons are simultaneously particles *and* waves. Not only that, but whether they behaved as particles or waves seemed to depend, unexplainably, on how the laboratory observer and his experimental apparatus interacted with them. Heisenberg developed what is known as the "uncertainty principle," which dictates that either the momentum or position of a subatomic, or quantum, entity can be known, but not both. This principle is still accepted today.[103]

Suddenly, Niels Bohr's model of the atom, electrons orbiting the atomic nucleus like planets, was outdated. The predictable orbits of electrons were replaced by "clouds" which specified only the *probability* of finding an electron in a given location. Einstein was dissatisfied with the probabilities and interpretations of quantum mechanics, saying that "God does not throw dice." He began the pursuit of a "Theory of Everything" that would return quantum mechanics to a deterministic science. He was not successful.

The observer's mysterious influence in the laboratory and loss of absolute predictability in quantum experiments generated intense interest in the scientific community. For some[104], it became a potential avenue of divine agency in the universe.[105]

***

As a man who has devoted his whole life to the most clear headed science, to the study of matter, I can tell you as a result of my research about atoms this much: There is no matter as such. All matter originates and exists only by virtue of a force which brings the particle of an atom to vibration and holds this most minute solar system of the atom together. We must assume behind this force the existence of a conscious and intelligent mind. This mind is the matrix of all matter. *Max Planck*[106]

***

Quantum mechanics is certainly imposing. But an inner voice tells me that it is not yet the real thing. The theory says a lot, but does not really bring us any closer to the secret of the "old one." I, at any rate, am convinced that He does not throw dice. *Albert Einstein*[107]

***

[To Einstein] Stop telling God what to do. *Niels Bohr.*[108]

***

If God has made the world a perfect mechanism, He has at least conceded so much to our imperfect intellects that in order to predict little parts of it, we need not solve innumerable differential equations, but can use dice with fair success. *Max Born*[109]

***

To Einstein's famous question expressing his abhorrence of quantum mechanics, "Does God throw dice?" the Judeo-Christian answer is not, as so many have wrongly supposed, a denial, but a very positive affirmative. For only in a world in which the laws of nature govern events in accordance with the casting of dice can the Bblical view of a world whose history is responsive to God's will prevail. *William G. Pollard*[110]

***

It was my science that drove me to the conclusion that the world is much more complicated than can be explained by science. It is only through the supernatural that I can understand the mystery of existence. *Allan Sandage*[111]

***

My faith begins with the realization that a supreme intelligence brought the universe into being and created man. *Arthur H. Compton*[112]

***

I believe that the universe was created and is sustained by some power that we call God. *John A. Russell*[113]

***

This most beautiful system of the sun, planets and comets, could only proceed from . . an intelligent and powerful Being. *Isaac Newton*[114]

***

I am personally persuaded that a super-intelligent Creator exists beyond and within the cosmos, and that the rich context of congeniality shown by our universe, permitting and encouraging the existence of self-conscious life, is part of the Creator's design and purpose. *Owen Gingerich*[115]

***

As far as natural science is concerned, advances in physics in this century have abolished the merely mechanical view of the universe which earlier centuries had espoused. Many detailed problems remain about the nature and location of God's activity, but its general conceivability has been enhanced by recent developments in science. Neither God nor man is perceived as caught in the grip of relentless causal rigidity. *John Polkinghorne*[116]

***

We can't understand the universe in any clear way without the supernatural. *Allan Sandage*[117]

***

There seems to be some force influencing the evolution of societies; to deny its existence is to deny any purpose to life. *John G. Phillips*[118]

***

Einstein was born into an Abrahamic faith; in his own view, he was deeply religious. Now this sense of wonder leads most scientists to a Superior Being—der Alte, the Old One, as Einstein affectionately called the Deity—a Superior Intelligence, the Lord of all Creation and Natural Law. *Abdus Salam*[119]

***

The great initial success of quantum theory cannot convert me to believe in that fundamental game of dice. ...I am absolutely convinced that one will eventually arrive at a theory in which the objects connected by laws are not probabilities but conceived facts. *Albert Einstein*[120]

***

The laws of physics could never have actually built the universe. Some agency must have been involved. *John Lennox*[121]

***

Many have a feeling that somehow intelligence must have been involved in the laws of the universe. *Charles H. Townes*[122]

***

Is the lingering notion which remains in the minds of some really a thought, that God could not just as easily by his word speak power into existence around centres, as He could first create nuclei and then clothe them with power? *Michael Farraday*[123]

***

From the intrinsic evidence of His creation, the Great Architect of the Universe now begins to appear as pure mathematician. *Sir James Jeans*[124]

***

We know that whenever God is experienced, it is an experience exactly as real as a direct sense impression, as real as one's own personality. As such He must be missing from the space-time picture. 'I do not meet with God in space and time', so says the honest scientific thinker, and for that reason he is reproached by those in whose catechism it is nevertheless stated: 'God is Spirit'. *Erwin Schrödinger*[125]

***

Anyone not shocked by quantum mechanics has not understood it. *Niels Bohr*[126]

***

I'm not an atheist and I don't think I can call myself a pantheist. The problem involved is too vast for our limited minds. We are in the position of a little child entering a huge library filled with books in many languages. The child knows someone must have written those books. It does not know how. It does not understand the languages in which they are written. The child dimly suspects a mysterious order in the arrangement of the books but doesn't know what that is. That, it seems to me, is the attitude of the most intelligent human toward God. *Albert Einstein*[127]

***

Max Planck

***

Anybody who has been seriously engaged in scientific work of any kind realizes that over the entrance to the gates of the temple of science are written the words: Ye must have faith. It is a quality which the scientist cannot dispense with. *Max Planck*[128]

***

I strongly believe in the existence of God, based on intuition, observations, logic, and also scientific knowledge. *Charles H. Townes*[129]

***

I am now convinced that theoretical physics is actually philosophy. *Max Born*[130]

***

I cannot seriously believe in [quantum theory] because ... physics should represent a reality in time and space, free from spooky action at a distance. *Albert Einstein*[131]

FIVE

LET THERE BE LIGHT: THE BIG BANG

And God said "Let there be light," and there was light.

Genesis 1:3

In addition to quantum mechanics, scientists of the 20th century made at least one other significant discovery that strengthened belief in the supernatural—the "Big Bang" theory. Up until then, the theory that the universe had an actual "beginning" had not been popular with scientists. [132] For many years, physicists such as Fred Hoyle found the idea of a beginning to be far too much like the Biblical account of creation in Genesis, and argued instead for a "steady state" universe—one that had no beginning.[133]

The Big Bang was first hypothesized by Belgian priest and astronomer Georges Lemaitre in 1927, but decades would pass before any "proof" could be found to support it. In 1948, Ralph Alpher and Robert Herman predicted that if the Big Bang had actually occurred there should be detectable background radiation. In the 1960's, American scientists Arno Penzias and George W. Wilson detected this background radiation.

Now widely accepted, the Big Bang Theory says that the entire universe—matter and energy—unexplainably burst forth from a single infinitesimally small point in space about 14 billion years ago. This event marked the beginning of time. Eventually the laws of physics, such as gravity, caused particles of matter to form into planets, stars, and galaxies.

Scientists who believed in creation and the Creator found their position greatly strengthened as a result of the Big Bang Theory's acceptance.[134] Evidence of the Big Bang's background radiation caused physicist George Smoot to comment, "If you're religious, it's like seeing God."[135]

***

The big bang creation of the universe is good news for God. It means there was a creation. *Gerald Schroeder*[136]

***

With respect to the Big Bang, astronomers [had previously] resisted the notion that the universe had a beginning, a singularity that smacks all too much in their philosophy of a miraculous event. Many were delighted, however by the congruence between the words of Genesis 1, "Let there be light!" and a hypothesized Big Bang in which the universe began with a mighty burst of energetic photons. *Owen Gingerich*[137]

***

First of all, we have this very solid conclusion that the universe had an origin, the Big Bang. Fifteen billion years ago, the universe began with an unimaginably bright flash of energy from an infinitesimally small point. That implies that before that, there was nothing. I can't imagine how nature, in this case the universe, could have created itself. And the very fact that the universe had a beginning implies that someone was able to begin it. And it seems to me that had to be outside of nature. And that sounds like God. *Francis Collins*[138]

***

There is no doubt that a parallel exists between the big bang as an event and the Christian notion of creation from nothing. *George Smoot*[139]

***

God created the universe out of nothing in an act which also brought time into existence. Recent discoveries, such as observations supporting the Big Bang and similar astronomical phenomena, are wholly compatible with this view. *Henry Margenau*[140]

***

Current research in astrophysics seems to indicate that the ultimate origin of the universe may be not only unknown but unknowable. That is, if we assume the Big Bang which present evidence strongly supports, there is no real way to find out what came before the Big Bang. It is surely right to pursue as far as possible the scientific understanding of the origins of the universe, but it is probably wrong to think that we have final answers and that there are no further surprises to come. From a religious point of view, we assume that God did it and hope to find out something about how he did it. *Arthur Schawlow*[141]

***

Subtle is the Lord, but malicious He is not. *Albert Einstein*[142]

***

The Big Bang seems to point strongly toward a Creator, since .. the question of what came before is left hanging in the air. *Francis Collins*[143]

***

Albert Einstein

***

Now we see how the astronomical evidence leads to a biblical view of the origin of the world. All the details differ, but the essential element in the astronomical and biblical accounts of Genesis is the same, the chain of events leading to man commenced suddenly and sharply, at a definite moment in time, in a flash of light and energy. *Robert Jastrow*[144]

***

Certainly there was something that set it off. Certainly, if you're religious, I can't think of a better theory [the Big

Bang] of the origin of the universe to match with Genesis. *Robert Wilson*[145]

***

In my mind a conflict between believing in God and believing in the Big Bang theory of the Universe never existed. In fact, just the opposite has occurred. The incredible beauty, wonder, and simplicity of the Universe have helped me to strengthen my belief in God and seek deeper knowledge about this relationship. I believe the Universe evolved by the laws of physics, but I also believe that it was God who originated these laws. *Giovanni Fazio*[146]

***

[God] solves some deeply troubling questions about what came before the Big Bang and why the universe seems to be tuned for us to be here. *Francis Collins*[147]

***

I do not understand how the scientific approach alone, as separated from a religious approach, can explain an origin of all things. It is true that physicists hope to look behind the "Big Bang" and possibly to explain the origin of our universe as, for example, a type of fluctuation. But then, of what is it a fluctuation and how did this in turn begin to exist? In my view, the question of origin seems always left unanswered if we explore from a scientific view alone. Thus, I believe there is a need for some religious or metaphysical explanation if we are to have one. *Charles H. Townes*[148]

***

Astronomy leads us to a unique event, a universe which was created out of nothing, one with the very delicate balance needed to provide exactly the conditions required to permit life, and one which has an underlying (one might say "supernatural") plan. Thus the observations of modern science seem to lead to the same conclusions as centuries-old intuition. At the same time, most of our modern scientific intuition seems to be more comfortable with the world as described by the science of yesterday. *Arno Penzias*[149]

***

The big bang, the most cataclysmic event we can imagine, on closer inspection appears finely orchestrated. *George Smoot*[150]

***

The God having the creative force to make the entire observable universe in a dense dot of pure energy is incomprehensible. *Owen Gingerich*[151]

***

There remains the question of how the Big Bang was initiated, but it seems unlikely that science will be able to elucidate this....the exquisite order displayed by our scientific understanding of the physical world calls for the divine. *Vera Kistiakowsky*[152]

***

The tendency of modern physics is to resolve the whole material universe into waves, and nothing but waves.... These concepts reduce the whole universe to a world of light, potential or existent, so that the whole story of its creation can be told with perfect accuracy and completeness in the six words: 'God said, Let there be light'. *Sir James Jeans*[153]

***

As to the first cause of the universe, in the context of expansion, this is left for the reader to insert, but our picture is incomplete without Him. [God] *Edward Milne*[154]

***

Science raises questions that it can never answer. Why did the big bang eventually lead to conscious beings who question the purpose of life and the existence of the universe? This is where religion is necessary. *Antony Hewish*[155]

OUR FINELY TUNED UNIVERSE

Would it not be strange if a Universe without purpose accidentally created humans who are so obsessed with purpose?

— *Sir John Templeton*[156]

In 1973, theoretical physicist Brandon Carter presented a paper that introduced what is today known in cosmology as the "anthropic principle." The anthropic principle is the philosophical argument that observations of the physical universe *must* be compatible with the conscious life that observes it. Today, scientists realize that there are at least 10 physical constants in the universe that, had any one of them failed to have the exact value they have, life as we know it would have never happened. The impact of slight variations in these constants might have been anything from the failure of atomic particles to form into stars to the loss of the biochemical conditions necessary for development of intelligent life.

Most atheists acknowledge these precise values are necessary for our existence, hence supporting what is known as the "weak" anthropic principle. But they balk at supporting the "strong" anthropic principle, which is more teleological in nature and states that the universe is in some sense *compelled* to have conscious life eventually emerge. Atheists accept the weak anthropic principle because they posit that our universe is only one of an infinite number of other universes that lack these precise values, and consequently it is only "dumb luck" that we exist in this one.

Believers in the Creator consider the infinite universes theory to be a *non sequitur* to the question of whether God created the universe. Infinite universes might address the advent of life, but it doesn't provide of the answer to "why is there something instead of nothing?"

***

God to me is a mystery—but is the explanation for the miracle of existence, why there is *something* instead of *nothing*. *Allan Sandage*[157]

***

A common sense interpretation of the facts suggests that a super-intellect has monkeyed with physics, as well as with chemistry and biology, and there are no blind forces worth speaking about in nature. The numbers one calculates from the facts seem to me so overwhelming as to put this conclusion almost beyond question.... Rather than accept the fantastically small probability of life having arisen through the blind forces of nature, it seemed better to suppose that the origin of life was a deliberate intellectual act. *Sir Fred Hoyle* [158]

***

I do not feel like an alien in the universe. The more I examine the universe and the details of its architecture, the more evidence I find that the universe in some sense must have known we were coming. *Freeman Dyson*[159]

***

When you look from the perspective of a scientist at the universe, it looks as if it knew we were coming. There are 15 constants – the gravitational constant, various constants about the strong and weak nuclear force, etc. – that have precise values. If any one of those constants was off by even one part in a million, or in some cases, by one part in a million million, the universe could not have actually come to the point where we see it. Matter would not have been able to coalesce; there would have been no galaxy, stars, planets or people. That's a phenomenally surprising observation. It seems almost impossible that we're here. And that does make you wonder – gosh, who was setting those constants anyway? Scientists have not been able to figure that out. *Francis Collins*[160]

***

What is clear is that life, as we know it, would not be possible if there were very small changes to either physics or the expanding universe that we see around us. There are many aspects of physics, which, if they were different, would prevent any life at all existing….We are now realizing that the universe is a very extraordinary place, in the sense that it is fine-tuned so that life will exist. *George Ellis*[161]

***

I believe in God. It makes no sense to me to assume that the Universe and our existence is just a cosmic accident. *Antony Hewish*[162]

***

The world is just so wonderful that I can't imagine it was just having come by pure chance. *Arthur Schawlow*[163]

***

Sir Arthur Eddington

***

It is impossible to trap modern physics into predicting anything with perfect determinism because it deals with probabilities from the outset. *Sir Arthur Eddington*[164]

***

God created everything by number, weight and measure. *Sir Isaac Newton*[165]

***

Overwhelmingly strong proofs of intelligent and benevolent design lie around us. *Lord Kelvin* [166]

***

I believe the book of nature .. suggests a God of purpose and a God of design. And I think my belief makes me no less a scientist. *Owen Gingerich*[167]

***

In fact we live in what appears to be a very finely tuned universe. The physical constants that describe how the forces of nature work with high quantitative accuracy are exactly right to allow life to exist and evolve and thrive for a meaningful length of time. Even tiny deviations from their measured values would have precluded life. One could try to explain this away by imagining that there could be a very large number of other universes, each with different fundamental constants and forces, so that this one that enables life as we know it is a statistical accident. If that were true, it would still be incredible that this "multi-verse" would be of such special character that even one universe within it would be a birthplace for life. *Jennifer Wiseman*[168]

***

I believe in God. In fact, I believe in a personal God who acts in and interacts with the creation. I believe that the observations about the orderliness of the physical universe, and the apparently exceptional fine-tuning of the conditions of the universe for the development of life suggest that an Intelligent Creator is responsible. *William Phillips*[169]

***

Order is manifestly maintained in the universe...governed by the sovereign will of God. *James P. Joule*[170]

***

I cannot believe that our existence in this universe is a mere quirk of fate, an accident of history, an incidental blip in the great cosmic drama. .. We are truly meant to be here. *Paul Davies*[171]

***

Dare a scientist believe in design? There is, I shall argue, no contradiction between holding a staunch belief in supernatural design and working as a creative scientist. *Owen Gingerich*[172]

***

It is the perfection of God's works that they are all done with the greatest simplicity. He is the God of order and not of confusion. *Sir Isaac Newton*[173]

***

As biochemists discover more and more about the awesome complexity of life, it is apparent that its chances of originating by accident are so minute that they can be completely ruled out. Life cannot have arisen by chance. *Sir Fred Hoyle*[174]

***

When you realize that the laws of nature must be incredibly finely tuned to produce the universe we see, that conspires to plant the idea that the universe did not just happen, but that there must be a purpose behind it. *John Polkinghorne*[175]

***

I think that God originated the universe and life. *Shoichi Yoshikawa*[176]

***

Whether atheist or theist a thoughtful person can only stand in awe of the way the universe seems designed as a home for humankind. *Owen Gingerich*[177]

***

There are at least 10 fundamental properties of the Universe that together have just the right values to make life possible. ...Thus we have the situation that if the overall density of matter in the Universe had been higher or lower than it was at the very beginning, and by an utterly infinitesimal percentage, we would either have a lifeless Universe with black holes instead of stars, or else nothing but a tenuous gas filling the Universe instead of stars and planets. A millionth of a millionth percent difference either way at the time of the Big Bang would have doomed the Universe one way or the other. That is an impressive fine tuning. *Bernard Haisch*[178]

***

The conception of an omniscient and omnipotent personal God, [w]ho made everything in accordance with a rational plan and purpose, contributed to the notion of a rationally structured creation. *Mehdi Golshani*[179]

***

To me, belief in a final cause, a Creator-God, gives a coherent understanding of why the universe seems so congenially designed for the existence of intelligent, self-reflective life. It would take only small changes in numerous physical constants to render the universe uninhabitable. Somehow, in the words of Freeman Dyson, this is a universe that knew we were coming. I do not claim that these considerations are proof for the existence of a Creator; I claim only that to me, the universe makes more sense with this understanding. *Owen Gingerich*[180]

DO MULTIPLE UNIVERSES MAKE SENSE?

Nothing can be created from nothing.

— *Lucretius*[181]

Scientists who do not countenance a supernatural cause for the universe have developed some fantastic cosmological theories[182] to avoid the Creator. Most of these, such as String Theory and Linde's Chaotic Inflationary Multiverse, involve the "many universes" hypotheses. There are major problems with each of these. First, they require a tremendous amount of "fine tuning" of their parameters in order to make their math work. Second, they cannot be proven or disproven via cosmological observation and experiment.[183] The inability to be proven by observation and fine tuning of parameters is similar to the theory that God created the universe, except in the latter case God did the fine tuning.[184]

Some of these theories say the universe has "always" existed, and there was no beginning. However, there are strong scientific reasons weighing against the theories of continuous or "cyclic" universes. First, the Second Law of Thermodynamics says that the universe is irreversibly "running down." Second, there is the Borde-Vilenkin-Guth Theorem, which was developed in 2003 by physicists Arvine Borde, Alex Vilenkin, and Alan Guth. It states all of the many universes theories still require a beginning—they cannot extend indefinitely into the past.[185]

Perhaps an *infinite* number of universes could explain the reason for our anthropocentric universe, and we have found ourselves in the bowl of porridge that was "just right." But which is more believable: faith in God, who supernaturally created our one universe out of nothing, or faith in an *infinite* number of universes that somehow created and re-created themselves out of *almost* nothing?[186]

***

The fine tuning of the universe provides *prima facie* evidence of deistic design. Take your choice: blind chance that requires multitudes of universes or design that requires only one. *Edward R. Harrison*[187]

***

I do not believe that the universe can create itself. Typical attempts to provide a scientific answer to the question "Can the universe create itself?" implicitly start with a "first" universe, out of which other universes are generated. It is here I would ask Andrei Linde [a supporter of multiverse theory] where his first universe came from that spawns his multiple baby universes *ad infinitum*. *Allan Sandage*[188]

***

Whoever wishes for a universe which can continue indefinitely in activity must lead a crusade against the Second Law of Thermodynamics. *Sir Arthur Eddington*[189]

***

Charles Townes

***

Some scientists argue that "well, there's an enormous number of universes and each one is a little different. This one just happened to turn out right." Well, that's a postulate, and it's a pretty fantastic postulate – it assumes there really are an enormous number of universes and that the laws could be different for each of them. The other possibility is that ours was planned, and that's why it has come out so specially. *Charles Townes*[190]

***

If the answer is simple, God is answering. *Albert Einstein*[191]

***

The doctrine of creation isn't about how things began, it's about why things exist, what holds the world in being. The Christian belief is that it is the will of God that holds the world in being. *John Polkinghorne*[192]

***

To say that there is no observable proof for such a macro-universe is an understatement. The laws of nature exclude the possibility of seeing outside our universe even if there is an outside. It is a theory that can never be tested by observation. *Gerald Schroeder*[193]

***

My conclusion is that the many-universes theory can at best explain only a limited range of features, and then only if one appends some metaphysical assumptions that seem no less extravagant than design. In the end, Occam's razor[194] compels me to put my money on design, but, as always in matters of metaphysics, the decision is largely a matter of taste rather than scientific judgment. *Paul Davies*[195]

***

To have a chain of contingent entities [universes] stretching back into the past, none of which explains its own existence and none of which causally depends on an entity or cause which explains its own existence, is simply unintelligible. *William Stoeger*[196]

The law that entropy always increases, holds, I think, the supreme position among the laws of Nature ..[so] if your theory is found to be against the second law of thermodynamics I can give you no hope; there is nothing for it but to collapse in deepest humiliation. *Sir Arthur Eddington*[197]

***

Truth is ever to be found in simplicity, and not in the multiplicity and confusion of things. *Sir Isaac Newton*[198]

***

There is no way to test the multiverse hypotheses. This is ruled out even in principle because with different laws of physics in different universes—the key hypotheses—there is no possibility of any observation. So it is not a scientifically provable theory. Thus belief in a multiverse is a faith, no matter how you dress it up in scientific language and mathematics. *Bernard Haisch*[199]

***

I have difficulty conceiving spontaneous creation in the distant past as a scientific process or accepting that matter has been in existence forever. *John A. Russell*[200]

***

Since the possibility of there being other universes is highly speculative, the most conservative and firm conclusion is that our universe is the only universe and it was designed for life. Which to my mind implies a Designer. *Gerald Schroeder*[201]

***

The multiverse theory just postpones the problem; the issue of ultimate causality remains…it just pushes the final question back one stage further. *George Ellis*[202]

***

Not even by supposing the eternity of the world can we escape the ultimate, extramundane reason of things, i.e., God. *Gottfried Leibniz*[203]

***

Reflections on the Creation of the universe lead to reflections about the Creator, which may surprise some readers to hear. It's also common to hear the opinion that science leads to atheism. This is an opinion that I can't share; I even find it somewhat absurd. *Walter Thirring*[204]

***

Laws which enable the universe to come into being spontaneously seem themselves to be the product of exceedingly ingenious design. *Paul Davies*[205]

***

To think that science already knows enough to be certain there are no mystical forces is illogical. *Charles H. Townes*[206]

EIGHT

HUMAN CONSCIOUSNESS AND THE MIND OF GOD

My thoughts are not your thoughts, nor are your ways my ways.

— *Isaiah 55:8*

One of the earliest mysteries of quantum mechanics was that it indicated the experimenter's act of conscious observation in the laboratory had an impact on the behavior of the quantum particles being observed. This phenomenon, usually referred to as the *quantum measurement problem*, generated great interest and controversy among the founders of quantum mechanics in the early twentieth century, with some saying the cause was supernatural.

Quantum theory, and the impact of the mind of the observer in quantum experiments, continues to baffle scientists. One of the most mysterious phenomena is known as "quantum entanglement," which occurs when two quantum particles that have interacted with each other can subsequently be found to have predictable, correlated behavior, even when they are separated by large distances.

Some neuroscientists and physicists believe the function of the human mind is governed by quantum theory. This is because, they say, the spaces between the neurons of the brain conduct signals using parts of atoms called ions, which function according to the rules of quantum, rather than classical, physics.[207] Some physicists believe quantum entanglement may play an important part in the brain's function as well, and could form the basis of an explanation of consciousness.

Some of the most famous scientists of history have broached the subject of "knowing the mind of God." What do they mean? Others discuss God as a "universal mind." How is human consciousness, morality, and free will influenced by God? These topics are rarely addressed by contemporary mainstream scientists.[208]

***

Quantum mechanics allows us to think of special divine action as 'the providential determination of otherwise undetermined events.' *Robert John Russell*[209]

***

We discover that the universe shows evidence of a designing or controlling power that has something in common with our own individual minds—not, so far as we have discovered, emotion, morality, or aesthetic appreciation, but the tendency to think in a way which, for want of a better word, we describe as mathematical. *Sir James Jeans.*[210]

***

As we have seen repeatedly, we cannot formulate any science without reference to the observer and, again as a matter of history, progress in fundamental science has been made by increasingly recognizing the role of the observer. It seems to me therefore that we cannot think about the Universe without the concept of personality. Cosmology requires, I venture to assert, the concepts of Creator and of personality, and together these mean God. *Sir William McCrea*[211]

***

I believe .. that neither physical science nor psychology can ever 'explain' human consciousness. To me, then, human consciousness lies outside science, and it is here that I seek the relationship between God and man. *Nevill Mott*[212]

***

A universal mind would be, I think, a fairly plausible inference from the present state of scientific theory. *Sir Arthur Eddington*[213]

***

My religiosity consists of a humble admiration of the infinitely superior spirit who reveals himself in the slight details we are able to perceive with our frail and feeble minds. That deeply emotional conviction of the presence of a superior reasoning power, which is revealed in the incomprehensible universe, forms my idea of God. *Albert Einstein*[214]

***

I regard consciousness as fundamental. I regard matter as derivative from consciousness. We cannot get behind consciousness. Everything that we talk about, everything that we regard as existing, postulates consciousness. *Max Planck*[215]

***

I maintain that the human mystery is incredibly demeaned by scientific reductionism, with its claim in promissory materialism to account eventually for all of the spiritual world in terms of patterns of neuronal activity. This belief must be classed as a superstition .. we have to recognize that we are spiritual beings with souls existing in a spiritual world as well as material beings with bodies and brains existing in a material world. *Sir John Eccles*[216]

***

The eternal mystery of the world is its comprehensibility...the fact it is comprehensible is a miracle. *Albert Einstein*[217]

***

It would not be surprising if it should turn out that the origin and destiny of the energy in the universe cannot be completely understood in isolation from the phenomena of life and consciousness....And the design of the inanimate universe may not be as detached from the potentialities of life and intelligence as scientists of the twentieth century have tended to suppose. *Freeman Dyson*[218]

***

I incline to the idealistic theory that consciousness is fundamental, and that the material universe is derivative from consciousness, not consciousness from the material universe... In general the universe seems to me to be nearer to a great thought than to a great machine. It may well be, it seems to me, that each individual consciousness ought to be compared to a brain-cell in a universal mind. *Sir James Jeans*[219]

***

Quantum mechanics is telling us that consciousness creates reality. Naturally this has profound consequences for the interpretation of our own nature, the seemingly fixed reality we take to be our universe, and, yes, even why it may make more sense to trace everything back to a conscious intelligence rather than simply inanimate fields and forces. *Bernard Haisch* [220]

***

If I consider reality as I experience it, the primary experience I have is of my own existence as a unique self-conscious being which I believe is God-created. *Sir John Eccles*[221]

***

As a scientist trained at the Massachusetts Institute of Technology I was convinced I had the information to exclude [God] from the grand scheme of life. But with each step forward in the unfolding mystery of the cosmos, a subtle yet pervading ingenuity, a contingency kept shining through, a contingency that joins all aspects of existence into a coherent unity. While this coherence does not prove the existence of a Designer, it does call out for interpretation. *Gerald Schroeder*[222]

***

The consciousness of the presence of God is the only guarantee of true self-knowledge. Everything else is mere fiction, fancy portraiture—done to please one's friends or self, or to exhibit one's moral discrimination at the expense of character. *James Clerk Maxwell*[223]

***

I want to know how God created this world. I am not interested in this or that phenomenon, in the spectrum of this or that element. I want to know his thoughts. The rest are details. *Albert Einstein*[224]

***

The doctrine that the world is made up of objects whose existence is independent of human consciousness turns out to be in conflict with quantum mechanics and with facts established by experiment. *Bernard d 'Espagnat*[225]

***

[An] aspect of God that I recognize is the basis of existence and of life and of motivation, which I think of as a conscious Power. This Power appears to me as having a special concern for its conscious creatures who share the responsibility for shaping their part of the world. *Arthur H. Compton*[226]

***

Erwin Schrodinger

***

Science cannot tell us a word about why music delights us, of why and how an old song can move us to tears. *Erwin Schrödinger*[227]

***

Are we merely part of the same picture as [the galaxies], or is it possible that we are part of the artist? Are they perchance only a dream, while we are the brain-cells in the mind of the dreamer? *Sir James Jean*[228]

***

As we have seen repeatedly, we cannot formulate any science without reference to the observer and, again as a matter of history, progress in fundamental science has been made by increasingly recognizing the role of the observer. It seems to me therefore that we cannot think about the Universe without the concept of personality. Cosmology requires, I venture to assert, the concepts of Creator and of personality, and together these mean God. *Sir William McCrea*[229]

***

I believe there is one "gap" for which there will never be a scientific explanation, and that is man's consciousness. *Nevill Mott*[230]

***

I hold that God is the totality of the universe; this includes all scientific principles, all matter and energy, and all life-forms. The existence of the universe requires me to conclude that God exists. *Robert A. Naumann*[231]

***

For me, the fundamental content of belief in God is that there is a Mind and Purpose behind the history of the universe. *John Polkinghorne*[232]

***

I draw the obvious conclusion that God also exists, and that my existence depends entirely upon Him at each and every moment. This conclusion is so evident that I am confident that the human mind can know nothing more evident or more certain. *René Descartes*[233]

***

An acquaintance with natural laws means no less than an acquaintance with the mind of God therein expressed. *James P. Joule*[234]

***

Since materialist solutions fail to account for our experienced uniqueness, we are constrained to attribute the uniqueness of the psyche or soul to a supernatural spiritual creation. *Sir John Eccles*[235]

***

We have learnt that the exploration of the external world by the methods of physical science leads not to a concrete reality but to a shadow world of symbols, beneath which those methods are unadapted for penetrating. Feeling that there must be more behind, we return to our starting point in human consciousness—the one centre where more might become known. *Sir Arthur Eddington*[236]

***

The very fact that the universe is creative, and the laws have permitted complex structures to emerge and to develop to the point of consciousness—in other words, that the universe has organized its own self-awareness—is for me powerful evidence that there is 'something going on" behind it all. The impression of design is overwhelming. *Paul Davies*[237]

***

I believe we ultimately come from a superior entity to which awe and respect is due and which we shouldn't try to approach by trying to conceptualize it too much. It's more a question of feeling. This I adhere to completely. *Bernard d'Espagnat*[238]

***

Consciousness cannot be accounted for in physical terms. For consciousness is absolutely fundamental. It cannot be accounted for in terms of anything else. *Erwin Schrödinger*[239]

NINE

THE SCIENCE OF
LIMITATIONS

There is a thought that stops all thought. That is the thought that ought to be stopped.

— G. K. Chesterton[240]

Some today believe that the mainstream science establishment has totally dismissed the "theory" of the existence of God. The usual criticism leveled towards belief in the Creator is that, unlike science, the existence of God is "un-provable."[241] As demonstrated herein, many major cosmological theories today are also un-provable.

The famous polymath, Michael Polanyi, has shown that the commonly held belief that scientific truth is always derived without bias, based solely upon experimental data, is a myth. In his primary work, Personal Knowledge: Towards a Post-Critical Philosophy, Polanyi concludes that absolute objectivity is a false ideal, because all knowledge claims, even those which are derived from rules, rely ultimately on personal judgment.

Blind prejudice can have a devastating effect in science—much as Einstein's bias against the concept of an expanding universe did. He was so prejudiced against the idea of an expanding universe that he added a factor Λ, called a cosmological constant, to his general relativity equations in order to force the math to result in a static universe. When astronomer Edwin Hubble later proved that the universe was expanding, Einstein lamented that inserting a cosmological constant had been "the biggest blunder of my life."[242]

Once scientists reach the point where they totally close their minds to a possible truth, such as the existence of a supernatural Creator, then British writer G. K. Chesterton's concern of "a thought that stops all thought" becomes reality.

***

As a society, we have made the mistake of thinking that because science can answer some questions very well, it might eventually be able to answer all questions. Scientists used to be quite modest in their claims. But recently a number of them have been growing more ambitious, as if the illusory power we have handed them has affected their judgment. The result has been a number of grandiose claims that can be neither justified nor fulfilled. *David Darling*[243]

***

The man of science should be essentially a rebel, a prophet rather than a priest, one who should not be ashamed of finding himself in opposition to the hierarchy. *Edward Arthur Milne*[244]

***

Today's dogma holds that matter is eternal. The dogma comes from the intuitive belief of people (including the majority of physicists) who don't want to accept the observational evidence that the universe was created – despite the fact that the creation of the universe is supported by all the observable data astronomy has produced so far. As a result, the people who reject the data can arguably be described as having a 'religious' belief that matter must be eternal. These people regard themselves as objective scientists. *Arno Penzias*[245]

***

There is no doubt that many scientists are opposed temperamentally to any form of metaphysical, let alone mystical arguments. They are scornful of the notion that there might exist a God, or even an impersonal creative principle or ground of being that would underpin reality and render its contingent aspects less starkly arbitrary. Personally I do not share their scorn. *Paul Davies*[246]

***

My opinion is that most people believe in intelligent design as a reasonable explanation of the universe, and this belief is entirely compatible with science. So it is unwise for scientists to make a big fight against the idea of intelligent design. The fight should be only for the freedom of teachers to teach science as they see fit, independent of political or religious control. It should be a fight for intellectual freedom, not a fight for science against religion. *Freeman Dyson*[247]

***

Steven Hawking ended his book *A Brief History of Time* with the statement that if this theory about the nature of the universe was upheld, it would help us "know the mind of God."[248] Hawking may be a genius, but his opinions about God carry no more weight than those of his next-door neighbor. *David Darling*[249]

***

There is a reluctance [in scientific circles] to reveal yourself as a believer, the opprobrium is so severe. *Allan Sandage*[250]

***

While many scientists ascribe to Theistic Evolution, they are in general reluctant to speak out for fear of negative reaction from their scientific peers, or perhaps from fear of criticism from the theological community. *Francis Collins*[251]

***

[Atheists] are creatures who–in their grudge against the traditional 'opium for the people'–cannot bear the music of the spheres. *Albert Einstein*[252]

***

Today physics is in a deep, deep slump. Thousands of papers are written every year on string theory, and yet there is still not one serious experimental test that might show that strings actually exist. *Bernard Haisch*[253]

***

Science may explain all the processes whereby the universe evolves its own destiny, but that still leaves room for there to be a meaning behind existence. *Paul Davies*[254]

***

According to Hawking,[255] the laws of physics, not the will of God, provide the real explanation as to how life on

Earth came into being. The Big Bang, he argues, was the inevitable consequence of these laws 'because there is a law such as gravity, the universe can and will create itself from nothing'....But, as both a scientist and a Christian, I would say that Hawking's claim is misguided. He asks us to choose between God and the laws of physics, as if they were necessarily in mutual conflict.... contrary to what Hawking claims, physical laws can never provide a complete explanation of the universe. Laws themselves do not create anything, they are merely a description of what happens under certain conditions. *John Lennox*[256]

***

Gravity explains the motions of the planets, but it cannot explain who set the planets in motion. God governs all things and knows all that is or can be done. *Sir Isaac Newton*[257]

***

The essential fact is simply that *all* the pictures which science now draws of nature, and which alone seem capable of according with observational fact, are mathematical pictures ... They are nothing more than pictures— fictions if you like, if by fiction you mean that science is not yet in contact with ultimate reality. *Sir James Jeans*[258]

***

God acts not in supposed gaps in the laws of nature, but rather in and through the laws themselves. They are expressions of God's creative activity in nature. *William Stoeger*[259]

***

To reject the explanation of an intelligence behind the origin of our Universe simply because one believes that there cannot conceivably be such an intelligence is really no different from faith in the equivalent of a holy book. In this case the faith is in reductionist materialism. Positing the existence of perhaps infinite other universes as a possible explanation is legitimate. But to argue that that must be true because the alternative of an intelligence just cannot be true is simply to worship at the altar of reductionist materialism. That is how the practice of science can morph into the faith of scientism. *Bernard Haisch*[260]

***

I think science can get a bad press by scientists appearing to be too arrogant in taking on more than perhaps they should, so it's as well to lace definitive statements with a certain amount of humility [regarding Stephen Hawking's claim[261] there is no need for a Creator]. *Paul Davies*[262]

***

Hawking's argument[263] appears ... illogical when he says the existence of gravity means the creation of the universe was inevitable. *John Lennox*[264]

***

There are metaphysical problems, which cannot be disposed of by declaring them meaningless. For, as I have repeatedly said, they are "beyond physics" indeed and demand an act of faith. We have to accept this fact to be honest. There are two objectionable types of believers: those who believe the incredible and those who believe that "belief" must be discarded and replaced by "the scientific method." *Max Born*[265]

***

The atheistic view that the universe just happens to be here without purpose and yet with exquisite logical structure appears to me to be obtuse. *Fred Hoyle*[266]

***

Materialists consider that we are explained entirely by combinations of small uninteresting things like atoms or quarks. With this notion of an ultimate reality that is unaccessable, it seems more natural to consider that we owe our existence to something that is higher than us. *Bernard d'Espagnat*[267]

***

Science is based on the assumption that the universe is thoroughly rational and logical at all levels. Miracles are not allowed. This implies that there should be reasons for the particular laws of nature that regulate the physical universe. Atheists claim that the laws exist reasonlessly and that the universe is ultimately absurd. As a scientist, I find this hard to accept. There must be an unchanging rational ground in which the logical, orderly nature of the universe is rooted. Is this rational ground like the timeless God of Augustine? Perhaps it is. *Paul Davies*[268]

***

Under these conditions it is no wonder, that the movement of atheists .. eagerly makes use of progressive scientific knowledge and in a presumed unity with it, expands in an ever faster pace its disintegrating action on all nations of the earth and on all social levels. I do not need to explain in any more detail that after its victory not only all the most precious treasures of our culture would vanish, but — which is even worse — also any prospects at a better future. *Max Planck*[269]

***

Scientists cannot bear the thought of a natural phenomenon which cannot be explained, even with unlimited time and money. There is a kind of religion in science, it is the religion of a person who believes there is order and harmony in the universe and every effect must have its cause, there is no first cause... This religious faith of the scientist is violated by the discovery that the world had a beginning under conditions in which the known laws of physics are not valid, and as a product of forces or circumstances we cannot discover. When that happens, the scientist has lost control. *Robert Jastrow*[270]

***

We can see that the principal content of creation *ex nihilo* is simply to underscore the insight that the existence of something, whether it be energy, material particles, or operative laws, requires a cause which either necessarily exists in itself, or ultimately rests on a cause which necessarily exists in itself—a primary cause, the first in the causal chain, not needing another cause to explain its existence. *William Stoeger*[271]

***

We should suspect an intention to reduce God to a system of differential equations. That fiasco at any rate [must be] avoided. ... We have learnt that the exploration of the external world by the methods of physical science leads not to a concrete reality but to a *shadow world of symbols*, beneath which those methods are unadapted for penetrating. *Sir Arthur Eddington*[272]

***

The strident atheists who grab the public limelight are not representative of all scientists. And they are not acting scientifically in any event, because they are asserting iron-clad belief in some-thing that cannot be proven: that there is no God. That is certainly a matter of faith. *Bernard Haisch*[273]

***

When some scientists try to say that science proves that God does not exist, that is an intellectually incorrect statement, a false philosophy. And that has been known since the time of Immanuel Kant. *George Ellis*[274]

***

We know nothing about it [God and the universe] at all. All our knowledge is but the knowledge of school children. Possibly we shall know a little more than we do now. But the real nature of things we shall never know, never. *Albert Einstein*[275]

***

It is time to begin a new and creative interaction between theology and science—an interaction which honors and respects the integrity of each partner, an interaction in which convictions are held self-critically and honest engagement is prized, an interaction which focuses specifically on the most rigorous theories of mainstream natural science and the most central positions of mainline theology. *Robert John Russell.*[276]

BIOGRAPHIES

Our knowledge is only a shadow in comparison with the greatness of God.

— *Immanuel Kant*[277]

I included in this book nearly 250 quotations from 70 scientists who support the concept of a creator God. Many, but not all, of the scientists quoted are believers in the God of the Bible. Many are Christian, some Jewish, and a few follow other major world religions. Some would not identify themselves with a particular religion. Some believe in a non-interfering God, a God of "harmony" like seventeenth century Dutch philosopher Baruch de Spinoza. Some would consider themselves theists, and others, simply deists. Others would not be able say exactly what they believe, but like Einstein would simply say, "I'm no atheist."

I tried to avoid quoting scientists who are openly atheistic, even when their own statements could be used against them to support the case for a Creator. When I found it useful to use quotes from scientists who are skeptics or openly atheistic, I put these in endnotes rather than in the body of the text.

Generally, the chapter quotes have come from scientists such as physicists, cosmologists, and astronomers—although a few mathematicians, chemists, and geneticists are included as well.

The reader can be assured that the scientists quoted herein are academically well qualified, even if not universally known, in their chosen fields. Although I did not elaborate on their education, all of the twentieth and twenty-first century scientists quoted have at least a PhD in a science field—usually physics or mathematics.

Francis Bacon (1561–1626)　English scientist and pioneer of the scientific method, which is sometimes referred to as the Baconian method. Bacon argued that valid knowledge must be empirically rooted in the natural world.

Niels Bohr (1885–1962)　Danish physicist and Nobel laureate for his investigation of the structure of atoms and the radiation emanating from them. He developed the theory of electrons orbiting the nucleus and the idea that an electron can drop from a higher energy orbit to a lower one, emitting a photon in the process. This became the basis for quantum theory. He and Einstein had a long-standing debate over the validity of quantum theory, which Bohr ultimately won.

Max Born (1882–1970)　German-born physicist and mathematician who was instrumental in the development of quantum mechanics. He also made contributions to solid-state physics and optics and, along with Walther Bothe, won the 1954 Nobel Prize in physics for the coincidence method.

Sir Robert Boyd (1922–2004)　British physicist and founding director of the Mullard Space Science Laboratory of the University College London. Pursued atmospheric physics research and his theses was on New Techniques for the Study of Ionized gases.

Robert Boyle (1627–1691)　Irish chemist, physicist, and theologian. Known for Boyle's Law, which describes the inversely proportional relationship between pressure and volume of a gas in a closed, constant temperature system. As a scientist and theologian, he rejected the popular view that the Bible was a scientific textbook and yet believed in the absolute harmony between scientific statements in the Bible and experimental science.

Sir William H. Bragg (1862–1942)   British physicist. Awarded the 1915 Nobel Prize in Physics for his contribution to the analysis of crystal structures by means of X-rays.

Francis Collins (born 1950)   American physician-geneticist, noted for his landmark discoveries of disease genes and his leadership of the Human Genome Project (HGP) and described by the Endocrine Society as "one of the most accomplished scientists of our time." He currently serves as Director of the National Institutes of Health in Bethesda, Maryland.

Arthur H. Compton (1892–1962)   American physicist and Nobel Prize laureate in physics for his discovery of the Compton Effect, which demonstrates the "particle" concept of electromagnetic radiation. Championed the position of human freedom based upon quantum indeterminacy.

Nicolaus Copernicus (1473–1543)   Mathematician and astronomer and first person to develop a comprehensive heliocentric cosmology, which displaced Earth from the center of the universe. His work *De revoutionibus orbium coelestium* is often considered the starting point of modern astronomy, beginning the scientific revolution.

David Darling (Born 1953)   English astronomer, lecturer, and science writer. Author of over 40 science books, as well as the *Internet Encyclopedia of Science.*

Paul Davies (Born 1946)   British-born theoretical physicist, cosmologist, astrobiologist and best-selling author. His subject of study has been diverse, including quantum fields in curved space-time, the cause of ripples in the cosmic afterglow after the big bang, the origin of the universe, and the origin of life. Winner of 1995 Templeton Prize.

René Descartes (1596–1650)   French mathematician and philosopher who was a key figure of the Scientific Revolution. Credited as the Father of analytic geometry. His *Meditations on First Philosophy* partially concerns theology. Known also for his famous "proof" of the existence of God, and as the originator of the statement "I think, therefore I am."

Freeman Dyson (born 1923)   British-born American theoretical physicist and mathematician, famous for his work in quantum field theory, solid-state physics, astronomy and nuclear engineering. Dyson is a member of the Board of Sponsors of the Bulletin of the Atomic Scientists. Dyson is best known for demonstrating in 1949 the equivalence of the formulations of quantum electrodynamics and the invention of the Dyson series. Winner of 2000 Templeton Prize.

Sir John Eccles (1903–1997)   Australian neurophysiologist that won the 1963 Nobel Prize in Physiology or Medicine for his work on the synapse of the peripheral nervous system. He was also key to a number of other important developments in neuroscience and a world-renown authority on brain function.

Sir Arthur Eddington (1882–1944)   British astrophysicist that was a theoretical researcher of the interior of stars and interpreter of the theory of relativity. The Eddington Limit, the natural limit to the luminosity of stars, is named in his honor. Along with Sir James Jeans, is considered a founder of British cosmology.

Albert Einstein (1879–1955)   German-born theoretical physicist who in 1916 developed the theory of general relativity, causing a revolution in physics. For this achievement, Einstein is often regarded as the father of modern physics, and one of the most prolific intellects in human history. He received the 1921 Nobel Prize in Physics "for his services to theoretical physics, and especially for his discovery of the law of the photoelectric effect". The latter was pivotal in establishing quantum theory within physics. Also well known for the development of his special theory of relativity in1905.

George Ellis (born 1939)   South African cosmologist and mathematician, considered one of the world's leading theorists in cosmology. Co-authored with Stephen Hawking the book *The Large Scale Structure of Space-Time* and winner of 2004 Templeton Prize.

Bernard d'Espagnat (born 1921)   French theoretical physicist and philosopher of science known for his research on the nature of reality and quantum mechanics. Known also for his leadership in the research on John

Bell's inequalities theorem, and its conflict with the predictions of quantum mechanics. Winner of 2009 Templeton Prize.

Michael Faraday (1791–1867) English physicist and chemist. A significant contributor to the fields of electro-magnetism and electrochemistry, he established the basis of electromagnetic field concept in physics, which was later enlarged upon by James Maxwell. In chemistry he discovered benzene, investigated clathrate hydrate of chlorine, and invented early forms of the Bunsen burner.

Giovanni Fazio Professor of Astronomy at Harvard-Smithsonian Center for Astrophysics. Research interests are study of star formation and evolution in nearby galaxies and ultra-luminous galaxies using infrared array cameras. Led the development of the Infrared Array Camera for NASA's Spitzer Space Telescope.

Ken Freeman Australian astronomer and astrophysicist. Currently professor of astronomy at the Research School of Astronomy and Astrophysics at Mount Stromlo Observatory of Australian University in Canberra. He specializes in research of the formation and dynamics of galaxies, globular clusters and dark matter.

Galileo Galilei (1564–1642) Italian mathematician, astronomer, and physicist who played a major role in the Scientific Revolution. His achievements include improvements to the telescope and consequent astronomical discoveries and support for Copernicus's heliocentric view of the universe. Sometimes called "the Father of Modern Science."

Owen Gingerich (born 1930)   Former Research Professor of Astronomy and the History of Science at Harvard University, and senior astronomer emeritus at the Smithsonian Astrophysical Observatory. In addition to his research and teaching, he has written many books on the history of astronomy. Influential member of the American Scientific Affiliation and a society of evangelical scientists.

Mehdi Golshani (1939)   Iranian theoretical physicist and philosopher. His main areas of research have been foundational and particle physics, physical cosmology, and philosophical implications of quantum mechanics.

Bernard Haisch   German-born American astrophysicist who has done research in solar-stellar astrophysics and stochastic electrodynamics. He has developed, along with Alfonso Rueda, a speculative theory that the non-zero lowest energy state of the vacuum, as predicted by quantum mechanics, might provide a physical explanation for the origin of inertia and be usable in spacecraft propulsion.

Edward R Harrison (1919–2007)   British astronomer and cosmologist, who spent much of his career at the University of Massachusetts and University of Arizona. He is noted for his work on the increase of fluctuations in the expanding universe, for his explanation of Olber's Paradox, and for his books on cosmology for lay readers.

Werner Heisenberg (1901–1976)   German theoretical physicist who made foundational contributions to quantum mechanics and is best known for asserting the uncertainty principle of quantum theory. Heisenberg, along with Max Born and P. Jordan, set forth the matrix formulation of quantum mechanics in 1925. Heisenberg was awarded the 1932 Nobel Prize in Physics for the development of quantum mechanics.

Sir Frederick William Herschel (1738–1822)   German-born British astronomer, William Herschel famous for the discovery of Uranus in addition to two of its major moons, Titania and Oberon. He also discovered two moons of Saturn and infrared radiation.

Sir John Frederick William Herschel   (1792–1871) English astronomer and mathematician, son of Sir William Herschel. Named seven of Saturn's moons and four of Uranus's moons. Originated the use of the Julian day system in astronomy. Investigated the chemical power of ultraviolet rays and color blindness. Made many significant contributions to the science of photography.

Antony Hewish (born 1924)   British radio astronomer who, together with Martin Ryle, won the Nobel Prize for Physics in 1974 for his work on the development of radio aperture synthesis and its role in the discovery of pulsars.

Sir Fred Hoyle (1915–2001)   English astronomer and mathematician noted primarily for his contribution to the theory of stellar nucleosynthesis and his initial rejection of the "Big Bang" theory. He and colleagues developed the Steady State Theory in 1948. During the course of his work he realized that the universe's chance of developing carbon-based life by accident was extremely unlikely. Hoyle, an atheist until that time, said that this suggestion of a guiding hand left him "greatly shaken."

Robert Jastrow (1925–2008)   American astronomer, physicist, and cosmologist. He was Chief of the Theoretical Division at NASA from 1958-1961, and became the founding director of NASA's Goddard Institute for Space Studies in 1961.

Sir James Jeans (1877–1946)   English physicist, astronomer, and mathematician. He made important contributions in many areas of physics, including quantum theory, the theory of radiation and stellar evolution. Along with Arthur Eddington is considered a founder of British cosmology. In 1928 was the first to conjecture a steady state cosmology, which was ruled out by the 1965 discovery of cosmic microwave background radiation from the Big Bang.

James P. Joule (1818–1889)   English physicist famous for his study of heat and its relationship to mechanical work. His research led to the theory of conservation of energy and the first law of thermodynamics. The SI unit of energy, the joule, is named after him. Also Joule's law, the relationship of heat generated by current flowing through a conductor.

Lord Kelvin (William Thomson) (1824–1907)   English mathematical physicist and engineer, who did important work in the mathematical analysis of electricity and formulation of the first and second laws of Thermodynamics. Also widely known for his realization that there was a lower limit to temperature, absolute zero, and absolute temperatures are stated in units of Kelvin in his honor.

Johannes Kepler (1571–1630)   German mathematician and astronomer. He is best known for Kepler's laws of planetary motion, which were based on empirical data obtained from Tycho Brahe's meticulous astronomical observations. Much of his work laid the groundwork for Isaac Newton's theory of universal gravitation.

Vera Kistiakowsky (born 1928)   American research physicist and educator. Author of more than 80 papers on nuclear physics, particle physics, and astrophysics. Elected a Fellow in the American Association and Advancement of Science in 1984.

Gottfried Leibniz (1646–1716)   German mathematician and philosopher. Made major contributions to physics, technology, logic, analytic philosophy, and developed differential and integral calculus independently of Newton. One of the early pioneers of the development of calculating machines. He is also known for his *Vindication of the Justice of God*.

John Lennox   Professor of Mathematics at the University of Oxford and Fellow in Mathematics and Philosophy of Science. His works include the mathematical *The Theory of Infinite Soluble Groups* and the religion-oriented *God's Undertaker – Has Science buried God?* He is also well known for his debates with various atheists such as Richard Dawkins on the existence of God.

McCrea, Sir William H. (1904–1999) English astronomer and mathematician. Using Albrecht Unsold's spectroscopic analysis of stellar atmospheres, in 1928 McCrea discovered that the Sun's composition is approximately three quarters hydrogen, one quarter helium, with 1% other elements.

Henry Margenau (1901–1997) German-U.S. physicist and philosopher of science. His work embraced investigation of intermolecular forces, spectroscopy, nuclear physics, electronics, and the relationship between quantum mechanics and human freedom. Developed a two stage model of free will in human decision making.

James Clerk Maxwell (1831–1879) Scottish physicist and mathematician. Developed the famous Maxwell's equations, demonstrating that electricity, magnetism, and light are all manifestations of the electromagnetic field. Demonstrated that electric and magnetic fields travel through space in the form of waves and the speed of light. His work laid the foundation for subsequent fields of special relativity and quantum mechanics.

Robert Millikan (1868–1953) American theoretical physicist and Nobel laureate in physics in 1923 for his work on measurement of the charge on the electron and for his work on the photoelectric effect. He verified experimentally Einstein's photoelectric equation, and made the first direct photoelectric determination of Planck's constant h.

Edward Arthur Milne (1896–1950)   British astrophysicist and mathematician who proposed the Milne model of the expanding universe as an alternative to Einstein's model. Milne's model is that of a sphere with homgeneous matter distribution within several billion light years of the center, which then increases to infinite density.

Sir Nevil Francis Mott (1905–1996)   English physicist and Nobel laureate in physics. He won the Nobel Prize in Physics in 1977 for his work on the electronic structure of magnetic and disordered systems, especially amorphous semiconductors. The award was shared with Philip Anderson.

Robert A. Naumann (born 1929)   American chemist and physicist. His areas of specialization are radioactivity, inorganic chemistry, nuclear physics, electromagnetic isotope separation radiochemical separation procedures, and examination of nuclear structure through radioactive and charged particle nuclear spectroscopy.

Isaac Newton (1643–1727)   English mathematician, physicist, and astronomer. He is regarded as one of the greatest scientists and mathematicians in history. In his *Philosophie Naturalis Principia Mathematica* he laid the foundations for classical mechanics, described universal gravitation and the three laws of motion. Among his many other accomplishments was the development of differential and integral calculus, independently of Leibniz.

Blaise Pascal (1623–1662)   French mathematician and physicist. Made important contributions to the study of fluids and did pioneering work on calculating machines.

Known for Pascal's law (physics), Pascal's theorem (math), and Pascal's Wager (theology). Defender of the scientific method.

Louis Pasteur (1822–1895) French chemist and microbiologist. Inventor of the pasteurization method, he also solved the mysteries of rabies, anthrax, chicken cholera, and silkworm diseases, and made significant contributions to the development of the first vaccines.

Arno Penzias (born 1933) American physicist, radio astronomer, and Nobel laureate in physics. Best known for his discovery, along with associate Robert Wilson, for cosmic microwave background radiation in 1964, which is considered a landmark verification of the Big Bang Theory. He and Robert Wilson received the 1978 Nobel Prize in physics for this work.

John G. Phillips (1917–2001) American astronomer and educator at University of California, Berkeley. His scientific contributions concerned the analysis of the spectra of molecules, and the design and construction of advanced instrumentation to speed and improve such analyses.

William Phillips (born 1948) American physicist and Nobel laureate in physics. In 1997 he won the Nobel Prize in physics for "development of methods to cool and trap atoms with laser light." He also won the 1998 Arthur Schawlow Prize in Laser Science.

Max Planck (1858–1947) German physicist who is regarded as the founder of quantum theory, for which he received the Nobel Prize in Physics in 1918. Did significant work on the problem of black body radiation, eventually developing what is known as the Planck Black Body Radiation Law. Known for Planck's postulate and Planck's constant. Also made significant contributions towards extension of Einstein's special theory of relativity.

John Polkinghorne (born 1930)   British theoretical physicist and Anglican priest. Former professor of Mathematical physics at the University of Cambridge, he is the author of many books on physics, as well as the relationship between science and religion. Winner of 2002 Templeton Prize.

Isidor Isaac Rabi (1898–1988)   Galician-born American physicist and Nobel laureate recognized in 1944 for his discovery of nuclear magnetic resonance. His investigations into the nature of the force binding protons to atomic nuclei eventually led to the creation of the molecular-beam magnetic-resonance (MRI) detection method.

John A. Russell (1913–2001)   American astronomer, educator, and pioneer in meteor spectroscopy. One of Russell's earliest spectra provided the first concrete evidence that the spectrum of a meteor depends primarily on the altitude and velocity of the meteoroid and not on its composition

Robert John Russell (born 1946)   American professor of physics and the founder/director of the Center for Theology and the Natural Sciences. Professor of Theology and Science in Residence at the Graduate Theological Union. Has authored and published various papers exploring consonance and dissonance between modern physics, evolutionary biology, and Christian theology.

Abdus Salam (1926–1996) Pakistani theoretical physicist and Nobel laureate in physics for his work on electroweak unification of the electromagnetic and weak forces. He shared the 1979 Nobel prize in physics with Steven Weinberg and Sheldon Glasgow for this work. Also known for the Pati-Salam model, magnetic photon, vector meson, Grand Unified Theory, and work on supersymmetry.

Allan Sandage (1926–2010) American astronomer. He was Staff Member Emeritus with the Carnegie Observatories in Pasadena, California. He is best known for, while working at the Palomar observatory, determining the first reasonably accurate value for the Hubble constant and the age of the universe. A leader in the verification of the theory of the expansion of the universe.

Arthur A. Schawlow (1921–1999) American physicist and Nobel laureate in physics. His research focused on optics, lasers, spectroscopy, superconductivity and nuclear resonance. He shared the 1981 Nobel Prize in Physics with Nicolas Bloembergen and Kai Siegbahn for their contributions to the development of laser spectroscopy.

Gerald Schroeder Israeli nuclear physicist, author and lecturer at *Aish Ha Torah's Discovery Seminar, Essentials and Fellowships* programs and Executive Learning Center, who focuses on what he perceives to be an inherent relationship between science and spirituality. Member of the United States Atomic Energy Commission.

Erwin Schrödinger (1887–1961)   Austrian physicist and theoretical biologist who is considered one of the fathers of quantum mechanics. He is also famous for a number of important contributions to physics, especially the Schrödinger equation, for which he received the Nobel Prize in Physics in 1933. In 1935, after extensive correspondence with his friend Albert Einstein, he proposed the famous Schrödinger's cat thought experiment.

George Smoot (born 1945)   American astrophysicist, cosmologist, and Nobel laureate in physics. He won the Nobel Prize in Physics in 2006 for his work on the Cosmic Background Explorer Satellite (COBE) with John C. Mather. COBE led to the measurement of the black body form and anisotropy of the cosmic microwave background radiation, further supporting the Big Bang theory.

William R. Stoeger   American astronomer and theologian. Staff scientist for the Vatican Observatory Research Group in Tucson, AZ, he specializes in theoretical cosmology, high-energy astrophysics, and interdisciplinary studies related to science, philosophy, and theology.

Joseph H. Taylor (born 1941)   American astronomer and Nobel laureate in physics. He and Russell Alan Hulse won the Nobel prize in physics in 1993 for their discovery of a "new type of pulsar, a discovery that has opened up new possibilities for the study of gravitation."

Walter Thirring (born 1927)   is an Austrian physicist after whom the Thirring model in quantum field theory is named. As a mathematical physicist and pioneer in quantum field theory, he worked with Einstein, Pauli, Heisenberg, and Schrödinger.

Charles H. Townes (born 1915)   American physicist, educator, and Nobel laureate in physics. Known for his work on the theory and application of the maser and other work in quantum electronics. He shared the Nobel Prize in Physics in 1964 with Nikolay Basov and Alexander Prokhorov. He is also known as the lead researcher in the construction of the Infrared Spatial Interferometer, and has a patent on maser technology. Winner of 2005 Templeton Prize.

Robert W. Wilson (born 1936)   American astronomer and Nobel laureate in physics. While working at Bell Labs with associate Arno Penzias, they discovered a source of noise in the atmosphere that they could not explain, finally identified as cosmic microwave background radiation. This provided important corroboration of the Big Bang theory. For this accomplishment he shared the 1978 Nobel Prize with Arno Penzias.

Jennifer Wiseman   Is an American astronomer and Chief of the Laboratory for Exoplanets and Stellar Astrophysics at NASA Goddard Space Flight Center. Wiseman is the Director for the American Association for the Advancement of Science's Dialogue on Science, Ethics, and Religion.

Shoichi Yoshikawa (1935–2010)   Japanese physicist and internationally known leader in fusion energy research. While at Princeton University, his areas of specialization and accomplishments were experimental and theoretical plasma physics and nuclear fusion device design.

ACKNOWLEDGEMENTS

First of all, I thank Jean, my wife of 33 years, for putting up with me. She has not only done that, but she has excelled in love, patience, kindness, joy, and the editing of manuscripts. I would like to thank Carol McCloud, author of *Have You Filled a Bucket Today?*, who inspired me to press on with this book when I had given up. My thanks also to friends that reviewed the manuscript and provided constructive criticism. These include Larry Henry, Leo Vadala, and Jim Linehan. Finally, I thank the folks at the Library of Congress's Prints and Photographs Division, who helped me with the images. These include Jeff Bridgers, Jon Eaker, Jan Grenci, Kristi Finefield, and Barbara Natanson.

PICTURE CREDITS

All pictures are courtesy of the Library of Congress with the exception of Johannes Kepler and Rene Descartes, which are in the Public Domain. The pictures of Arthur Compton and Erwin Schrodinger are from the Library of Congress's NYWTS Collection.

Front cover image of the Butterfly nebula is courtesy of NASA Space Telescope Science Institute.

END NOTES

1   Dascal, Marcelo (2008). "Lebniz on Creation: A contribution to his philosophical theology" by Daniel J. Cook, in *Lebniz: What Kind of Rationalist*, New York: Springer, p. 452.

2   In order to explain the existence of the universe, physicists always have to start with *something*, rather than nothing. Lawrence Krauss, in his recent book *A Universe from Nothing,* is no different. Krauss's "nothing" that he starts with includes the laws of physics and electromagnetic and/ or gravitational forces. Theoretical gravitational physicist Don N. Page of the University of Alberta says in an amazon book review of Krauss's book "To me as a fellow scientist, it appears Krauss has .. swung far wide of the goal, striking only the air with his philosophical speculations that do not address the truly deep questions of existence." Similarly, Alan Guth, who developed inflationary theory, says that to create the universe one would have to start with the laws of physics and some quantum gravity. He admits that currently physicists "haven't any clue" where the laws of physics came from. See Robert Lawrence Kuhn interview on this subject with Alan Guth at www.closertotruth.com.

3   Einstein, Albert (1970). *Out of My Later Years*, Westport, CT: Greenwood Press, p. 25.

4   People that believe that truth is attainable *only* through the pursuit of science, even to the exclusion of all other viewpoints, are supporters of what is typically called "scientism." For excellent

critiques of scientism and the arguments of the new atheists, see Alister and Joanna McGrath's *The Dawkins Delusion* (2007) and Ian Hutchinson's *Monopolizing Knowledge* (2011). An excellent critique of Richard Dawkins' *The God Delusion* is presented by agnostic H. Alan Orr in The New York Review of Books, which can be accessed at http://www.nybooks.com/articles/archives/2007/jan/11/a-mission-to-convert/?page=2

5   Stannard, Russell. (2000). *God for the 21ˢᵗ Century*. Radnor, PA: Templeton Foundation Press, p.165.

6   Brock & Freeman. (2008). *Shrouds of the Night*. New York: Springer, p. 369.

7   Richardson, & Slack. (2001). *Faith in Science, Scientists Search for Truth*. New York: Routledge, p. 28.

8   Calaprice, & Einstein. (2000). *Expanded Quotable Einstein*. Princeton, NJ: Princeton University Press, p. 213.

9   Planck, Max. (1932) "Where is Science Going?" Cited 12/22/11 in Wikiquote at http://en.wikiquote.org/wiki/Max_Planck,

10 Dyson, Freeman. (2000). *Progress in religion: A talk by Freeman Dyson*. Retrieved September 15, 2011 from http://www.uncommondescent.com/intelligent-design/interview-with-freeman-dyson/ .

11 Schrödinger, Erwin. (1996).*'Nature and the Greeks' and 'Science and Humanism.'* Cambridge: University Press, pp. 95-96.

12 Davies, Paul. (2000). *What happened before the Big Bang?* Accessed 9/9/11 from http://www.fortunecity.com/emachines/e11/86/big-bang.html.

13 Collins, Francis S. (2006). *The Language of God*, New York: Free Press, p. 88.

14 Barbour, Ian. (1997). *Religion and Science, Gifford Lecture Series*. New York: HarperOne, p. 198.

15 Russell, John A."I have difficulty accepting that matter has been in existence forever." From Margenau, & Varghees, eds. (1991). *Cosmos, Bios, Theos: Scientists Reflect on Science, God, and*

the Origins of the Universe, Life, and Homo Sapiens. Peru, IL: Open Court, p. 90.

16 Wiseman, Jennifer. (2007). *Science as an Instrument of Worship.* From BioLogos Foundation, accessed 9/15/11 from http:// biologos.org/uploads/projects/wiseman_white_paper.pdf.

17 *Teaching about Evolution and Science, Working Group on Teaching Evolution.* (1998) National Academy of Sciences, ISBN: 0-309-53221-3, p. 58.

18 Shawlow, Arthur. "One must ask why and not just how." From Margenau, & Varghees, eds. (1991). *Cosmos, Bios, Theos: Scientists Reflect on Science, God, and the Origins of the Universe, Life, and Homo Sapiens.* Peru, IL: Open Court, p. 105.

19 Jastrow, Robert. "What forces filled the universe with energy fifteen billion years ago?" From Margenau,& Varghees, eds. (1991). *Cosmos, Bios, Theos: Scientists Reflect on Science, God, and the Origins of the Universe, Life, and Homo Sapiens.* Peru, IL: Open Court, p. 46.

20 Margenau, Henry. "The laws of nature are created by God." From Margenau, & Varghees, eds. (1991). *Cosmos, Bios, Theos: Scientists Reflect on Science, God, and the Origins of the Universe, Life, and Homo Sapiens.* Peru, IL: Open Court, p. 62.

21 Collins, Francis S. (2006). *The Language of God*, New York: Free Press, p. 6.

22 Ellis, George. (2006) Quoted by Marty Morrissey at *Touching the limits of knowledge: Cosmology and our view of the world.* Accessed 12/19/11 at http://espg.sr.unh.edu/preceptorial/ Summaries_2006/summary15_MMor_2006.html

23 Smoot & Davidson. (2007). *Wrinkles in Time: Witness to the Birth of the Universe.* New York: Harper Perennial, p. 291.

24 Davies, Paul. (1993). *The Mind of God: The Scientific Basis for a Rational World.* New York: Simon & Schuster, p. 226

25 Ellis, George "On Rationality and Emotion, Faith and Hope in Being Human in a Scientific Age". Accessed 12/19/11 at http://members.fortunecity.es/lamb1/emotion.html

26 Schrödinger, Erwin. (1996).*'Nature and the Greeks' and 'Science and Humanism.'* Cambridge: University Press, p. 95.

27 Jeans, Sir James (1942) *Physics and Philosophy*, reprint 2003 by Kessinger Publishing, Whitefish, MT, p. 15.

28 Jastrow, Robert. (1992).*God and the Astronomers.* London: WW Norton, Readers Library, p.106.

29 Jammer, Max. (2011). *Einstein and Religion: Physics and Theology*, Princeton University Press, p. 91.

30 Sinologist Joseph Needham and physicist John D. Barrow have theorized, for instance, that belief in a divine being that "underwrote scientific enterprise" is what caused the scientists of western civilization to make more rapid advances in astronomy, physics, and cosmology than their counterparts in the East, who lacked a corresponding philosophy. Others have theorized that it was the Eastern civilization's belief in endless cycles of history—in which the world and its civilizations are repeatedly created, destroyed, and created again—that caused a sense of hopelessness and indifference to scientific investigation.

31 Pritchard, Duncan (2009). *What is this thing called knowledge?* New York: Routledge, p. 44.

32 Gingerich, Owen. (2006). *God's Universe.* New York: Belknap Press, p. 115.

33 Collins, Francis S. (2006). *The Language of God*, New York: Free Press, pp. 230-231.

34 Jaki, S. (1978). *The Road of Science and the Ways to God*, Edinburgh: Scottish Academic Press, p. 179.

35 Little, Eliakim and Robert S. (1845). *The Living Age*, Boston, MA: Waite, Peirce, and Co., Volume 5, p. 383.

36 Shawlow, Arthur. "One must ask why and not just how." From Margenau,& Varghees, eds. (1991). *Cosmos, Bios, Theos: Scientists Reflect on Science, God, and the Origins of the Universe, Life, and Homo Sapiens.* Peru, IL: Open Court, p. 106.

37 Begley, Sharon. 1998. "Science Finds God". *Newsweek*. Vol. CXXXII, No. 4, 27 July, pp. 44-49. U.S. Edition. New York, p. 47.

38 Block, D. and Freeman, K. (2008). *Shrouds of the Night*, New York, NY: Springer, p. 364.

39 Gladstone, John Hall. (2010). *Michael Faraday, 1874*. Memphis, TN: General Books, p59.

40 Stannard, Russell. (2000). *God for the 21st Century*. Radnor, PA: Templeton Foundation Press, p.165.

41 Newton, Sir. Isaac. (1848) *"Philosophiae Naturalis Principia Mathematica, Book III."* New York: Daniel Adee, p. 546.

42 Morris, Henry M. (1990). *Men of Science—Men of God*, Cajan, CA: Masters Books, pp. 29-30.

43 Cottingham, Stoothoff, & Murdoch, translators. (1984) *Philosophical Writings of Descartes, Volume 2,* p. 45. Cambridge: Cambridge University Press.

44 Schawlow, Arthur L. "One must ask why and not just how." From Margenau, & Varghees, eds. (1991). *Cosmos, Bios, Theos: Scientists Reflect on Science, God, and the Origins of the Universe, Life, and Homo Sapiens*. Peru, IL: Open Court, p. 107.

45 Noble, David (1999). *The Religion of Technology: The Divinity of Man and the Spirit of Invention,* New York: Penguin, p. 116.

46 W. Buchanan, John M. Baxter, and Alex Moir. (1910). *Selected Exhortations Delivered to Various Churches of Christ by the Late Michael Faraday*. Bridlington: John Leng & Co, p. 18

47 Campbell and Garnett, (1882), *The Life of James Clerk Maxwell*, Macmillan, p. 158.

48 Boyle, Robert. (1725). *The Philosophical Works of the Honourable Robert Boyle Esq: Abridged, Methodized, and Disposed Under the General Heads of Physics, Statics, Pneumatics, Natural History, Chymistry, and Medicine*, Innys, W. and J. Innys Publishers, p.149.

49 Phillips, William D. (2002). "Science and the Spiritual Quest II" (April 20; Interdisciplinary University of Paris)

from *Ordinary Faith, Ordinary Science*. A public lecture at the conference. UNESCO World Headquarters, Salle 1, Paris.

50 McGrath, Alistere E. (2012). *Mere Apologetics: How to help seekers and skeptics find faith*. Grand *Rapids, MI: Baker Books, p. 109.*

51 Collins, Francis S. (2011). Accessed 11/19/11 at http://scienceasworship.hostoi.com/index.html

52 Penzias, Arno.(1978) From New York Times interview on March 12, 1978. Cited in Bergman, Jerry. 1994. "Arno A. Penzias: Astrophysicist, Nobel Laureate," in *Perspectives on Science and Christian Faith, 46*.3: 183-187, (9/1994). The Journal of the American Scientific Affiliation, p. 183.

53 Buxhoeveden, D. and Woloschak, G. (2011). Science and the Eastern Orthodox Church. Surrey, UK: Ashgate, p. 133..

54 Compton, Arthur H. (1936). Quoted in *Chicago Daily News* (Magazine Section, Issue of April 12). Chicago, USA.

55 Hewish, Anthony. Cited in *Candid Science IV: Conversations with Famous Physicists* by Istvan Hargittai, London, Imperial College Press, 2004, 637.

56 Walsch, James Joseph. (1907). *Makers of Modern Medicine.* New York: Fordham University Press, p. 318.

57 Boyd, Sir Robert. Quoted in Poole, Mike. (1997). *God and the Scientists*. Worthing: CPO, p. . ISBN 1-901796-02-7.

58 Campbell and Garnett, (1882), *The Life of James Clerk Maxwell*, Macmillan, p. 426.

59 Brian VanDeMark, (2003). *Pandora's Keepers: Nine Men and the Atomic Bomb*, Little Brown & Co., Ch. 1).

60 Jeans, Sir James (1951). *The Growth of physical Science*, Cambridge: Cambridge University Press, p.173.

61 Gingerich, Owen. (2006). *God's Universe*. New York: Belknap Press, p. 9.

62 Ione, Amy. (2002). *Nature Exposed to our Method of Questioning*, Diatrope Press, p. 103.

63 Heisenberg, Werner. (1971). *Physics and Beyond*, New York: Harper & Row, pp. 82-83.

64 Collins, Francis S. (2006). *The Language of God*, New York: Free Press, p. 198.

65 Planck, Max. (Reprinted 1977). *Where is Science Going?* .New York, NY: W.W. Norton.

66 Calaprice, & Einstein. (2000). *Expanded Quotable Einstein*. Princeton, NJ: Princeton University Press, p. 214.

67 Heisenberg, Werner. Cited in *50 Nobel Laureates and other great scientists who believe in God,* by Dimitrov Tihomir.(1995-2008). An eBook, retrieved on 9/12/11 from http://nobelists.net/, p.13.

68 Naumann, Robert A. "Religion and science both proceed from acts of faith." From Margenau, & Varghees, eds. (1991). *Cosmos, Bios, Theos: Scientists Reflect on Science, God, and the Origins of the Universe, Life, and Homo Sapiens*. Peru, IL: Open Court, p. 71.

69 Phillips, John G. "Science asks what and how, while religion asks why." From Margenau,& Varghees, eds. (1991). *Cosmos, Bios, Theos: Scientists Reflect on Science, God, and the Origins of the Universe, Life, and Homo Sapiens*. Peru, IL: Open Court, p. 84.

70 Easterbrook, Gregg. (1997). "Science and God: A Warming Trend?" ("Of Lasers and Prayer"). *Science*. Vol. 277: 890-893; August 15. AAAS.

71 Journal of Chemical Education, v. 3. (1926). American Chemical Society, Section of Chemical Education.

72 Collins, Francis S. (2011). Accessed 11/19/2011 from http://www.pbs.org/faithandreason/transcript/coll-frame.html

73 Planck, Max. Religion and Natural Science (Lecture given in 1937). From *A Scientific Autobiography and Other Papers* (1949), translated by F. Gaynor, New York, p. 184.

74 Kistiakowsky, Vera. "The exquisite order of the physical world calls for the Divine." From Margenau,& Varghees, eds. (1991). *Cosmos, Bios, Theos: Scientists Reflect on Science, God, and the Origins of the Universe, Life, and Homo Sapiens*. Peru, IL: Open Court, p. 51.

75 Townes Charles H. Cited in *50 Nobel Laureates and other great scientists who believe in God,* by Dimitrov Tihomir. (1995-2008). An eBook, retrieved 9/12/11 from http://nobelists.net/, p. 16.

76 Polkinghorne, John. (2007). *Quantum Physics and Theology,* New Haven, CT: Yale University Press, p. 1.

77 Heisenberg, Werner. 1974. *Across the Frontiers.* (World Perspectives, Vol. 48). New York & San Francisco: Harper and Row Publishers, p. 213.

78 Leahy, Michael Patrick. (2007). *Letter to atheist.* Thompson Station, TN: Harpeth River Press, p. 61.

79 Bacon, Francis. (1875). *The Essays of Lord Bacon.* London: Longman and Green, Co., p. 64.

80 Block, D. and Freeman, K. (2008). *Shrouds of the Night,* New York: Springer, p. 371.

81 Smith, Alfred Emanuel (1903) *New Outlook,* v. 74, May-August. New York: The Outlook Company, p. 505.

82 Polkinghorne, John. (2007). *Quantum Physics and Theology,* New Haven, CT: Yale University Press, p. 8.

83 Planck, Max. (Reprinted 1977). *Where is Science Going?* .New York: W.W. Norton, p. 168.

84 Jacks, Hicks, Spinks, Gerard, and Short (1951). *The Hibbert Journal: A Quarterly Review of Religion, Theology, and Philosophy.* Hibbert Trust, v. 50-51, p. 198.

85 Schroeder, Gerald L. (2009). *The Science of God: The Convergence of Scientific and Biblical Wisdom.* New York: Free Press, p. 8.

86 Caroe, Gwendolen M. (1979). *William Henry Bragg (1862 - 1942): Man and Scientist.* Cambridge: Cambridge University Press, p. 171.

87 Johnson, Marjorie, ed. (1967). *The Cosmos of Arthur Holly Compton.* Introduction by Vannevar Bush. New York: Alfred A. Knopf, Inc., p. 374.

88 Millikan, Robert A. (1925). "A Scientist's God," in *Collier's, The National Weekly.* October 24. Collier's Publishing Company.

89   Phillips, William D. (2002). A letter to the compiler T. Dimitrov. May 19, 2002. Cited in *50 Nobel Laureates and other great scientists who believe in God,* by Dimitrov Tihomir. An eBook, retrieved on 9/12/11 from http://nobelists.net/, p. 19.

90   Taylor, Joseph H. (1995-2008). Cited in *50 Nobel Laureates and other great scientists who believe in God,* by Dimitrov Tihomir. An eBook, retrieved on September 12, 2011 from http://nobelists.net/ , p. 30.

91   Mott, Sir Nevill. "Science will never give us the answers to all our questions." From Margenau,& Varghees, eds. (1991). *Cosmos, Bios, Theos: Scientists Reflect on Science, God, and the Origins of the Universe, Life, and Homo Sapiens.* Peru, IL: Open Court, p. 64.

92   Heisenberg, Werner. (1971). *Physics and Beyond,* New York: Harper & Row, pp. 82-83.

93   Thompson, Silvanus Phillips. (1910). *The Life of Lord Kelvin, part 2,* London: Macmillan, p. 1103

94   Calaprice, & Einstein. (2000). *Expanded Quotable Einstein.* Princeton, NJ: Princeton University Press, p. 204.

95   Davies, Paul (1993). *The Mind of God,* New York: Touchstone, p. 161.

96   Salam, Abdus. (1976). "An interview with Dr. Robert Walgate" of *New Scientist.* (26 August 1976).

97   Rabi, Isidor Isaac. (1999). Quoted in *Physics Today* (Sept. 1999): p. 38.

98   Gashani, Mehdi "Theistic Science," quoted in Stannard, Russell. (2000). *God for the 21ˢᵗ Century.* Radnor, PA: Templeton Foundation Press, p.192.

99   Holton, Gerald. (1998) *The Advancement of Science and its Burdens.* Cambridge, MA: Harvard University Press, p. 91.

100   Jastrow, Robert. (1992). *God and the Astronomers.* London: WW Norton, Readers Library, p.107.

101 Chesterfield, Lord. Accessed 2/3/12 from http://www.
     quotesarchive.com/authors/c/lord-chesterfield/quotes/in-
     seeking-wisdom-thou-art-wise-in-ima .

102 Quantum theory was founded primarily by Max Planck.
     Werner Heisenberg led development of the observational
     and experimental aspects of the new theory, often called
     quantum mechanics.

103 Deterministic science was dealt another blow in the mid-
     dle of the twentieth century with the formalization of
     what is known today as deterministic chaos theory. Chaos
     theory studies the behavior of dynamical systems that are
     highly sensitive to initial conditions. Small differences in
     initial conditions can yield widely diverging outcomes for
     chaotic systems, making long-term prediction of future
     events impossible.

104 Per Nicholas Saunders' *Divine Action and Modern Science*
     (2002) scientists and scholars such as William G. Pollard,
     Arthur Compton, J. J. Thomson, George Thomson, E.
     T. Whittaker, Karl Heim, and Eric Mascall are among
     those believing, to varying degrees, that God influences
     the world thru quantum mechanics, chaos theory, and/or
     quantum chaos. See also John Jefferson Davis' *The Frontiers
     of Science and Faith* (2002).

105 As biochemist Arthur Peacocke said in his book *Creation
     and the World of Science*, p. 132: "How can God act in a
     world in which every event is tied to every other by
     regularities which the sciences explain with increasing
     power and accuracy?" Also, see W. G. Pollard, *Chance and
     Providence* (1958), articles by Murphy, Tracey, Russell, and
     Peacock in Robert J. Russell's (1995) *Chaos and Complexity*
     and John Polkinghorne (1998) *Belief in God in an Age of
     Science*, chapter three for a discussion of scientists' theories
     on God's actions through quantum uncertainty and chaos
     theory.

106 Planck, Max. "Das Wesen der Materie " [The Nature of Matter], speech at Florence, Italy (1944). From Archiv zur Geschichte der Max-Planck-Gesellschaft, Abt. Va, Rep. 11 Planck, Nr. 1797.

107 Calaprice, & Einstein. (2000). *Expanded Quotable Einstein.* Princeton, NJ: Princeton University Press, p. 215.

108 Bohr, Niels. Quoted from Wikiquote. Accessed 11/15/11 from http://en.wikiquote.org/wiki/Niels_Bohr

109 Schilpp, Paul A., ed. (1951). *Albert Einstein: Philosopher-Scientist.* Chicago, IL: Open Court, p. 176.

110 Pollard, William G. (1958) *Chance and Providence: God's Action in a World Governed by Scientific Law.* New York: Scribner's, p. 97.

111 Begley, Sharon. (1998). *Science Finds God.* Accessed 11/15/11 from http://www.washingtonpost.com/wp-srv/newsweek/science_of_god/scienceofgod.htm

112 Huxley & Luskin. (2004). *Evidence of Design of the Universe through Anthropic Principles.* Accessed 9/15/11 from http://www.ideacenter.org/contentmgr/showdetails.php/id/837

113 Russell, John A. "I have difficulty accepting that matter has been in existence forever." From Margenau, & Varghees, eds. (1991). *Cosmos, Bios, Theos: Scientists Reflect on Science, God, and the Origins of the Universe, Life, and Homo Sapiens.* Peru, IL: Open Court, p. 91.

114 Newton, Sir. Isaac. "Philosophiae Naturalis Principia Mathematica, Book III." Cited in Thayer, H. S. (1953). *Newton's Philosophy of Nature: Selections from his writings.* New York, NY: Hafner Library of Classics, p. 42.

115 Gingerich, Owen. (2006). *God's Universe.* New York: Belknap Press, p. 39

116 Polkinghorne, John. (2005). Science and Providence: God's Interactions With the World. Templeton Press, p. 18.

117 Sandage, Alan. Personal interview with F. Heeren, May 9, 1996. Cited in Heeren, Fred, (1998) *Show Me God.* Wheeling, IL: Day Star, p. 224.

118 Phillips, John G. "The exquisite order of the physical world calls for the Divine." From Margenau,& Varghees, eds. (1991). *Cosmos, Bios, Theos: Scientists Reflect on Science, God, and the Origins of the Universe, Life, and Homo Sapiens.* Peru, IL: Open Court, p. 85.

119 Salam, Abdus. "Science and religion: Reflections on transcendence and secularization." From Margenau, & Varghees, eds. (1991). *Cosmos, Bios, Theos: Scientists Reflect on Science, God, and the Origins of the Universe, Life, and Homo Sapiens.* Peru, IL: Open Court, p. 94.

120 Barbour, Ian G. (2000). *When Science Meets Religion: Enemies, Strangers, or Partners?* New York, NY: HarperOne, p. 67.

121 Lennox, John. (2010). *As a scientist I'm certain Stephen Hawking is wrong. You can't explain the universe without God.* Accessed 9/15/11 from http://www.dailymail.co.uk/debate/article-1308599/Stephen-Hawking-wrong-You-explain-universe-God.html.

122 Begley, Sharon. (1998). *Science Finds God.* Accessed 9/15/11 from http://www.washingtonpost.com/wp-srv/newsweek/science_of_god/scienceofgod.htm .

123 Russell, Colin A. (1985). *Cross-currents: Interactions Between Science and Faith,* Leicester: Inter-Varsity Press, p. 261.

124 Jeans, James (1939). *The Mysterious Universe,* Cambridge: Cambridge University Press, p. 134.

125 Schrödinger, Erwin. (1958). *Mind and Matter,* Cambridge: Cambridge University Press, p. 68.

126 Rosenblum, B. and Kuttner, F. (2011). *Quantum Enigma: Physics Encounters Consciousness.* New York: Oxford University Press, p. 113.

127 Isaacson, W. (2008). *Einstein: His Life and Universe.* New York: Simon and Schuster, p. 386.

128 Planck, Max (1932). Accessed 9/15/11 from Wikiquote at http://en.wikiquote.org/wiki/Max_Planck

129 Townes, Charles H. (1995-2008). Cited in *50 Nobel Laureates and other great scientists who believe in God*, by Dimitrov Tihomir. An eBook, retrieved on September 12, 2011 from http://nobelists.net/ , p. 16.

130 Born, Max. (1963). Quoted in *Schrödinger : Life and Thought* (1989) by Walter J. Moore. Cambridge: Cambridge University Press, p. 1.

131 Rosenblum and Kuttner (2011). *Quantum Enigma: Physics Encounters Consciousness*, Oxford University Press, p. 3.

132 The comment by Alexander Vilenkin illustrates scientists' dissatisfaction with the concept of a beginning, "For most physicists this [a beginning] was too much to take. A singular jump-starting of the universe looked like a divine intervention, for which they thought there should be no place in physical theory." (Vilenkin, *Many Worlds in One: The Search for Other Universes*, 2006, p. 24). See also Kathy Ferguson's discussion in Chapter 3 of her book *The Fire in the Equations: Science, Religion, and the Search for* God (1994).

133 Fred Hoyle may have never changed his mind about the Big Bang, but he later became an avid supporter of the concept of a Creator God.

134 Alex Vilenkin said "in the 1950's the accumulating evidence for the big bang inspired enthusiasm in theological circles and among some religiously inclined scientists." See *Many Worlds in One: The Search for Other Universes*, 2006, p. 176.

135 Smoot, George (1992). Quote at Lawrence Berkeley National Laboratory news conference, accessed 9/17/12 from http://physicsworld.com/cws/article/news/32559

136 Schroeder, Gerald L. (2009). *The Science of God: The Convergence of Scientific and Biblical Wisdom*. New York: Free Press, p. 216.

137   Gingerich, Owen (2006). *God's Universe*, Cambridge, MA: Harvard University Press, p. 48.

138   Collins, Francis S. (2006). Retrieved 11/19/11 from http://www.salon.com/2006/08/07/collins_6/

139   Smoot & Davidson. (2007). *Wrinkles in Time: Witness to the Birth of the Universe*. New York: Harper Perennial, p. 17?

140   Margenau, Henry. "The laws of nature are created by God." From Margenau, & Varghees, eds. (1991). *Cosmos, Bios, Theos: Scientists Reflect on Science, God, and the Origins of the Universe, Life, and Homo Sapiens*. Peru, IL: Open Court, p. 62.

141   Schawlow, Arthur L. "One must ask why and not just how." From Margenau, & Varghees, eds. (1991). *Cosmos, Bios, Theos: Scientists Reflect on Science, God, and the Origins of the Universe, Life, and Homo Sapiens*. Peru, IL: Open Court, p. 106.

142   Clark, R. W. (1974). *Einstein, the Life and Times*. New York: Harper Collins-Avon, p. 38.

143   Collins, Francis S. (2006). *The Language of God*, New York: Free Press, p. 77.

144   Jastrow, Robert. (1992).*God and the Astronomers*. London: WW Norton, Readers Library, p.14.

145   Wilson, George. Personal interview with F. Heeren, May 18, 1994. Cited in Heeren, Fred, (1998) *Show Me God*. Wheeling, IL: Day Star, p. 157.

146   Block, D. and Freeman, K. (2008). *Shrouds of the Night*, New York: Springer, p. 367.

147   Collins, Francis S. (2006). *The Language of God*, New York: Free Press, p. 81.

148   Townes, Charles. "The question of origin seems unanswered if we explore from a scientific view alone." From Margenau, & Varghees, eds. (1991). *Cosmos, Bios, Theos: Scientists Reflect on Science, God, and the Origins of the Universe, Life, and Homo Sapiens*. Peru, IL: Open Court, p. 123.

149 Penzias, Arno. "Creation is supported by all the data so far." From Margenau, & Varghees, eds. (1991). *Cosmos, Bios, Theos: Scientists Reflect on Science, God, and the Origins of the Universe, Life, and Homo Sapiens.* Peru, IL: Open Court, p. 83.

150 Smoot & Davidson. (2007). *Wrinkles in Time: Witness to the Birth of the Universe.* New York: Harper Perennial, p. 135.

151 Gingerich, Owen (2006). *God's Universe,* Cambridge, MA: Harvard University Press, p. 104.

152 Kistiakowsky, Vera. "The exquisite order of the physical universe calls for the Divine." From Margenau,& Varghees, eds. (1991). *Cosmos, Bios, Theos: Scientists Reflect on Science, God, and the Origins of the Universe, Life, and Homo Sapiens.* Peru, IL: Open Court, p. 52.

153 Jeans, James. (2010 reprint) *The Mysterious Universe,* Whitefish, MT: Kessinger Publishing, p. 69.

154 Vilenkin, Alex. (2006) *Many Worlds in One: The Search for Other Universes,* New York: Hill and Wang, p. 176.

155 Hewish, Antony. (1995-2008). A letter to the compiler T. Dimitrov, cited in *50 Nobel Laureates and other great scientists who believe in God,* by Dimitrov Tihomir. An eBook, retrieved on September 12, 2011 from http://nobelists. net/, pp. 29-30.

156 Block, D. and Freeman, K. (2008). *Shrouds of the Night,* New York, NY: Springer, p. 373.

157 Willford, J.N. March 12, 1991. *Sizing up the Cosmos: An Astronomers Quest.* New York Times, p. B9.

158 Hoyle, Fred (1981). Engineering and Science, November 1981. Cited by Owen Gingerich, *Let There Be Light: Modern Cosmogony and Biblical Creation, The World Treasury of Physics, Astronomy, and Mathematics,* ed. By Timothy Ferris. Boston, MA: Little, Brown, and Company, 1991) p. 393.

159 Dyson, Freeman. (1979). *Disturbing the Universe.* New York: Harper & Row, p. 250.

160 Collins, Francis S. (2006). Accessed 11/19/11 from http://www.salon.com/2006/08/07/collins_6/

161 Haisch, Bernard. (2010). *The Purpose-Guided Universe: Believing in Einstein, Darwin, and God.* Franklin Lakes, NJ: Career Press, p. 195.

162 Hewish, Antony. (1995-2008). A letter to the compiler T. Dimitrov, cited in *50 Nobel Laureates and other great scientists who believe in God,* An eBook by Tihomir, Dimitrov. Retrieved on September 12, 2011 from http://nobelists.net/ , p. 29.

163 Schawlow, Arthur L. 1998. *Optics and Laser Spectroscopy, Bell Telephone Laboratories, 1951-1961, and Stanford University Since 1961*, an oral history conducted in 1996 by Suzanne B. Riess (Interviewer/Editor), Regional Oral History Office, The Bancroft Library, University of California, Berkeley. Chapter 1, Part 5.

164 Eddington, Arthur. (1935). *New Pathways in Science.* Cambridge: At the University Press, p. 105.

165 Mackay, Alan L. (1991). *A Dictionary of Scientific Quotations,* Oxford: Taylor & Francis, p. 182.

166 Thompson, Silvanus Phillips. (1910). *The Life of Lord Kelvin, part 2*, London: Macmillan, p. 608.

167 Gingerich, Owen (2006). *God's Universe*, Cambridge, MA: Harvard University Press, p. 79.

168 Wiseman, Jennifer. (2007). "Science as an Instrument of Worship". From BioLogos Foundation, accessed 12/23/11 from http://biologos.org/uploads/projects/wiseman_white_paper.pdf.

169 Phillips, William D. (2002). A letter to the compiler T. Dimitrov. May 19, 2002. Cited in *50 Nobel Laureates and other great scientists who believe in God,* by Dimitrov Tihomir. An eBook, retrieved on September 12, 2011 from http://nobelists.net/, p. 19.

170 Steffens, H. J. (1979). *James Prescott Joule and the Concept of Energy*, Folkestone: Dawson Publishing, p. 142.

171 Davies, Paul (1993). *The Mind of God*, New York: Touchstone, p. 232.

172 Block, D. and Freeman, K. (2008). *Shrouds of the Night*, New York: Springer, p. 364.

173 Newton, Sir Isaac. Cited in *Rules for methodizing the Apocalypse*, Rule 9, from a manuscript published in *The Religion of Isaac Newton* (1974) by Frank E. Manuel, Oxford, UK: Oxford University Press, p. 120.

174 Hoyle, Fred, (1983). *The Intelligent Universe*, London: Michael Joseph, p. 12.

175 Begley, Sharon. (1998). *Science Finds God*. Retrieved on September 15, 2011 from http://www.washingtonpost. com/wp-srv/newsweek/science_of_god/scienceofgod.htm.

176 Yoshikawa, Shoichi. "The hidden variables of quantum mechanics are under God's power." From Margenau, & Varghees, eds. (1991). *Cosmos, Bios, Theos: Scientists Reflect on Science, God, and the Origins of the Universe, Life, and Homo Sapiens*. Peru, IL: Open Court, p. 135.

177 Gingerich, Owen (2006). *God's Universe*, Cambridge, MA: Harvard University Press, p. 40.

178 Haisch, Bernard. (2010). *The Purpose-Guided Universe: Believing in Einstein, Darwin, and God.* Franklin Lakes, NJ: Career Press, p. 70.

179 Golshani, Mehdi. Accessed 1/17/2012 from Wikipedia, at http://en.wikipedia.org/wiki/Mehdi_Golshani

180 Gingerich, Owen (2006). *God's Universe*, Cambridge, MA: Harvard University Press, p. 12.

181 Lucretius. From Wikiquote accessed 2/3/12 at http:// en.wikiquote.org/wiki/Lucretius .

182 For instance, physicist Alexander Vilenkin's cosmological theory says that every possible chain of events, no matter how improbable, has actually happened or is currently happening somewhere in the universe—an infinite number of times. Your own birth, for instance, has been replicated countless times, and your life has been lived with

various outcomes. See Alex Vilenkin's *Many worlds in One: The Search for Other Universes*, 2006.

183 See discussion by Robert J. Spitzer in *New Proofs for the Existence of God: Contributions of Contemporary Physics and Philosophy,* Grand Rapids, MI: Eerdman's, Chapter One.

184 In his paper, *The Matrix as Metaphysics*, former Director of Consciousness Studies at the University of Arizona David Chalmers made an interesting observation with his comment: "I don't know whether the Creation Hypothesis is true. But I don't know for certain that it is false. The hypothesis is clearly coherent, and I cannot conclusively rule it out." From http://consc.net/papers/matrix.html accessed 1/28/2012.

185 Alex Vilenkin said "It is said that an argument is what convinces reasonable men and a proof is what it takes to convince even an unreasonable man. With the proof [BVG Theorem] now in place, cosmologists can no longer hide behind the possibility of a past-eternal universe. There is no escape, they have to face the problem of a cosmic beginning. *Many Worlds in One: The Search for Other Universes*, 2006, p. 176.

186 Physicist Alan Guth says that science can explain how the universe can be created from *almost* nothing. All you need to start with are the laws of physics and a handful of "repulsive gravity." Of course, how one gets the laws and repulsive gravity out of *nothing* is a big problem. See personal interview with Alan Guth at http://www.closertotruth.com/video-profile/Why-is-There-Anything-At-All-Alan-Guth-/860 accessed 12/15/11.

187 Harrison, E. (1985). *Masks of the Universe*. New York, Collier Books, Macmillan, pp. 252, 263

188 Richardson, Mark W. (2003). *Science and the Spiritual Quest: New Essays by Leading Scientists,* Florence, KY: Psychology Press, Taylor & Francis Grp, p. 59.

189   Russell, Colin A. (1985). *Cross-currents: Interactions Between Science and Faith*, Leicester: Inter-Varsity Press, p. 219.

190   Townes, Charles. From *UC Berkely* interview, cited in http://www.adherents.com/people/pt/Charles_Hard_Townes.html, accessed 12/23/11.

191   Einstein, Albert. Accessed 12/27/11 from Wikiquote at http://en.wikiquote.org/wiki/Talk:Albert_Einstein.

192   Polkinghorne, John interview. Quoted in The BioLogos Forum's Videoblog presentation "From the Dust: Evolutionary Creation." Accessed 1/24/12 at http://biologos.org/blog/a-leap-of-truth

193   Schroeder, Gerald L. (2009). *The Science of God: The Convergence of Scientific and Biblical Wisdom*. New York: Free Press, p. 216.

194   According to Wikipedia, Occam's razor is "a principle that generally recommends that, from among competing hypotheses, selecting the one that makes the fewest new assumptions usually provides the correct one, and that the simplest explanation will be the most plausible until evidence is presented to prove it false."

195   Davies, Paul (1993). *The Mind of God*, New York: Touchstone, p. 220.

196   Stoeger, William. "The Origin of the Universe in Science and Religion." From Margenau, & Varghees, eds. (1991). Cosmos, Bios, Theos: Scientists Reflect on Science, God, and the Origins of the Universe, Life, and Homo Sapiens. Peru, IL: Open Court, p. 257.

197   Eddington, Sir Arthur (1923). *The Nature of the Physical World*, reproduced by Nabu Press, (2011), Charleston, SC: Nabu, p. 74.

198   Powell, Corey S. (2002). *God in the Equation: How Einstein Transformed Religion,* New York: Simon and Schuster, Free Press, p. 29.

199 Haisch, Bernard. (2010). *The Purpose-Guided Universe: Believing in Einstein, Darwin, and God.* Franklin Lakes, NJ: Career Press, p. 197.

200 Russell, John A. "I have difficulty accepting that matter has been in existence forever." From Margenau, & Varghees, eds. (1991). *Cosmos, Bios, Theos: Scientists Reflect on Science, God, and the Origins of the Universe, Life, and Homo Sapiens.* Peru, IL: Open Court, p. 90.

201 Schroeder, Gerald L. (2009). *The Science of God: The Convergence of Scientific and Biblical Wisdom.* New York: Free Press, p. 216

202 Haisch, Bernard. (2010). *The Purpose-Guided Universe: Believing in Einstein, Darwin, and God.* Franklin Lakes, NJ: Career Press, p. 197.

203 Russell, Bertrand. (2008). *A Critical Exposition of the Philosophy of Leibniz,* New York, NY: Cosimo, p.285.

204 Thirring, Walter. *Cosmic Impressions: Traces of God in the Laws of Nature,* Philadelphia: Templeton Press, pp. vii-viii.

205 Davies, Paul (1984). *Superforce,* New York: Simon and Schuster, p. 243.

206 Stannard, Russell. (2000). *God for the 21ˢᵗ Century.* Radnor, PA: Templeton Foundation Press, p.34.

207 See Mario Beauregard, *The Spiritual Brain: A Neuroscientist's Case for the Existence of the Soul* (2007). According to Beauregard, "Consciousness and other aspects of the mind, which can influence neural events, can occur independently of the brain, generally through aspects of quantum mechanics, "p. 107. In addition to quantum theory, emergence has been proposed by various scholars as a scientific explanation of the influence of the mind of God on human consciousness. For example in an interview on Closer to Truth, Philip Clayton said "If you can't give a mechanicstic account of thought .. then it's possible to say, though I can't explain the mechanisms, that there is a divine influence or lure on human thought which doesn't break natu-

ral law." Accessed 3/17/2012 at http://www.closertotruth. com/video-profile/Does-God-Intervene-in-the-World-Philip-Clayton-/1193 . See also Philip Clayton and Paul Davies, eds:*The Re-Emergence of Emergence: The Emergentist Hypotheses from Science to Religion (2008).*

208 See for instance George Ellis and Nancey Murphy (1996) *On the Moral Nature of the Universe: Theology, Cosmology,* and Ethics for an interesting discussion on God's interaction with mankind and subjects such as morality, cosmology, and kenoticism. See also John Polkinghorne's *Science and Providence: God's Interactions with the World (2005)*; and Nicholaus Saunders,*Divine Action and Modern Science* (2003).

209 Russell, Robert John. "Special Providence and Genetic Mutation: A New Defense of Theistic Evolution," quoted in Miller, Keith B., ed. (2003). *Perspectives on an Evolving Creation.* Grand Rapids, MI: Eerdmans's, p. 363.

210 Jeans, James (1939). *The Mysterious Universe*, Cambridge: Cambridge University Press, p. 188.

211 McRea, William H. "Models, Laws, and the Universe." From Wolfgang and Breck, eds. (1977) *Cosmology, History, and Theology.* New York: Plenum, p. 72

212 Mott, Nevill. (1997). *Can Scientists Believe? Some Examples of the Attitude of Scientists to Religion.* Oxford: Taylor and Francis, p. 8.

213 Raymond J. Seeger. "Eddington, Mystic Seeker," in *The Journal of the American Scientific Affiliation*, 36 (March 1984): page 36. Accessed 9/26/11 from http://www.asa3. org/ASA/PSCF/1984/JASA3-84Seeger2.html

214 Dukas and Hoffmann, (1981). *Albert Einstein: The Human Side.* Princeton, NJ: Princeton University Press, p. 66.

215 de Purucker, Gottfried. (1940). *The Esoteric Tradition.* California: Theosophical University Press.

216 Block, D. and Freeman, K. (2008). *Shrouds of the Night*, New York: Springer, p. 368.

217 Calaprice, & Einstein. (2000). *Expanded Quotable Einstein.* Princeton, NJ: Princeton University Press, p. 278.

218 Haisch, Bernard. (2010). *The Purpose-Guided Universe: Believing in Einstein, Darwin, and God.* Franklin Lakes, NJ: Career Press, p. 185.

219 Jeans, Sir James. Personal interview published in *The Observer*, London England. Accessed 1/19/2012 from http://en.wikipedia.org/wiki/Sir_James_Jeans

220 Haisch, Bernard. (2010). *The Purpose-Guided Universe: Believing in Einstein, Darwin, and God.* Franklin Lakes, NJ: Career Press, p. 66.

221 Eccles, Sir John. "A Divine Design: Some Questions on Origins." From Margenau, & Varghees, eds. (1991). *Cosmos, Bios, Theos: Scientists Reflect on Science, God, and the Origins of the Universe, Life, and Homo Sapiens.* Peru, IL: Open Court, p. 161.

222 Schroeder, Gerald L. (2009). *The Science of God: The Convergence of Scientific and Biblical Wisdom.* New York: Free Press, p. 26.

223 Maxwell, James Clerk letter to Rev. L. Campbell, Jan 31, 1858, quoted in Campbell & Garnett, (1882). *The Life of James Clerk Maxwell.* London: Macmillan, p. 150. Accessed 9/9/11 from http://www.sonnetsoftware.com/bio/maxbio.pdf,

224 Calaprice, & Einstein. (2000). *Expanded Quotable Einstein.* Princeton, NJ: Princeton University Press, p. 202.

225 Bernard, (1979). The Quantum Theory and Reality, Scientific American, Nov. 1979. Accessed 1/19/12 at http://www.scientificamerican.com/media/pdf/197911_0158.pdf

226 Compton, Arthur H. (1956) *Atomic Quest: A Personal Narrative,* Cambridge, MA: Oxford University Press, p. 345.

227 Schrödinger, Erwin. (1996).'*Nature and the Greeks' and 'Science and Humanism.'* Cambridge: University Press, p. 97.

228 Russell, John A. "I have difficulty accepting that matter has been in existence forever." From Margenau, & Varghees, eds. (1991). Cosmos, Bios, Theos: Scientists Reflect on Science, God, and the Origins of the Universe, Life, and Homo Sapiens. Peru, IL: Open Court, p. 92.

229 Block, D. and Freeman, K. (2008). *Shrouds of the Night*, New York: Springer, p. 366.

230 Mott, Neville. "Science will never give us the answers to all our questions." From Margenau,& Varghees, eds. (1991). *Cosmos, Bios, Theos: Scientists Reflect on Science, God, and the Origins of the Universe, Life, and Homo Sapiens*. Peru, IL: Open Court, p. 66.

231 Naumann, Robert A.. "Religion and science both proceed from acts of faith." From Margenau,& Varghees, eds. (1991). *Cosmos, Bios, Theos: Scientists Reflect on Science, God, and the Origins of the Universe, Life, and Homo Sapiens*. Peru, IL: Open Court, p. 72.

232 Polkinghorne, John. (2003). *Belief in God in an Age of Science*, New Haven, CT: Yale University Press, p. 1.

233 Descartes, Renee. (1993). *Meditations on First Philosophy* (third edition). Indianapolis, IN: Hackett Publishing, p. 36.

234 J.P. Joule, in a paper found with his scientific notebooks, as cited in: J.G. Crowther, *British Scientists of the Nineteenth Century*, Routledge & Kegan Paul, London, 1962, p. 139

235 Eccles and Robinson. (1985) *The Wonder of Being Human*, Boston, MA: Shambhala, p. 37.

236 Eddington, Sir Arthur (1929). *Science and the Unseen World*. New York: Macmillan.

237 Davies, Paul (2004). *The Cosmic Blueprint*, Conshohocken, PA: Templeton Press, p. 203.

238 d'Espagnat, Bernard. (2009) Personal interview at http://blogs.reuters.com/faithworld/2009/03/17/the-scientist-who-leaves-room-for-spirituality/ accessed 1/19/12.

239 Schrödinger, Erwin. (1984). "General Scientific and Popular Papers," in *Collected Papers,* Vol. 4. Vienna: Austrian Academy of Sciences. Friedr. Vieweg & Sohn, Braunschweig/Wiesbaden., p. 334.

240 Chesterton and Colson (2009). *Orthodoxy,* Chicago, IL: Moody Publishers, p. 54.

241 See Ferguson, Kathy (1994). *The Fire in the Equations: Science Religion, and the Search for God,* Chapter 3 for an interesting discussion on the "almost objective" position of science and academia.

242 Interest in the cosmological constant is coming back. Some tie it to the mysterious"dark energy that fills the universe."

243 Haisch, Bernard (2010) *The Purpose-Guided Universe: Believing in Einstein, Darwin, and* God. Franklin Lakes, NJ: Career Press, pp. 171-172.

244 Milne, Edward (1952). *Modern Cosmology and the Christian Idea of God,* Isle of Wight: Clarendon Press, pp. 8-9.

245 Penzias, Arno. As quoted in Bergman, Jerry. 1994. "Arno A. Penzias: Astrophysicist, Nobel Laureate," in *Perspectives on Science and Christian Faith,* 46.3: 183-187, (9/1994). The Journal of the American Scientific Affiliation, p. 183.

246 Davies, Paul (1993). *The Mind of God,* New York: Touchstone, p. 231.

247 Dyson, Freeman. Response to Christopher Morbey, cited at http://www.uncommondescent.com/intelligent-design/freeman-dyson-comments-on-id/, accessed 12/23/11.

248 Referring to Hawking's closing statement of his book *A Brief History of Time,* where Hawking said: "If we find the answer to that [a unified theory] it would be the ultimate triumph of human reason—for then we would know the mind of God."

249 Haisch, Bernard (2010). *The Purpose-Guided Universe: Believing in Einstein, Darwin, and God* Franklin Lakes, NJ: Career Press, p. 172.

250 Begley, Sharon. 1998. "Science Finds God". *Newsweek.* Vol. CXXXII, No. 4, 27 July, pp. 44-49. U.S. Edition. New York, p. 47.

251 Collins, Francis S. (2006). *The Language of God*, New York: Free Press, p. 202.

252 Calaprice, & Einstein. (2000). *Expanded Quotable Einstein.* Princeton, NJ: Princeton University Press, p. 214.

253 Haisch, Bernard. (2010). *The Purpose-Guided Universe: Believing in Einstein, Darwin, and God.* Franklin Lakes, NJ: Career Press, p. 175.

254 Davies, Paul (2004). *The Cosmic Blueprint*, Conshohocken, PA: Templeton Press, p. 203.

255 Reference is to Stephen Hawking's statement that because we have laws such as gravity, "the universe can and will create itself from nothing" and there "is no place for a creator," in various interviews and his book *The Grand Design* (2010).

256 Lennox, John. (2010). *As a scientist I'm certain Stephen Hawking is wrong. You can't explain the universe without God.* Accessed 9/15/11 from http://www.dailymail.co.uk/ debate/article-1308599/Stephen-Hawking-wrong-You-explain-universe-God.html.

257 From http://en.wikipedia.org/wiki/Isaac_Newton's_religious_views, accessed 12/3/11

258 Jeans, James (1939). *The Mysterious Universe*, Cambridge: Cambridge University Press, p. 111.

259 Stoeger, William, "Can God really act in our world and in our lives?" quoted in Stannard, Russell. (2000). *God for the 21ˢᵗ Century*. Radnor, PA: Templeton Foundation Press, p.167.

260 Haisch, Bernard. (2010). *The Purpose-Guided Universe: Believing in Einstein, Darwin, and God.* Franklin Lakes, NJ: Career Press, p. 43.

261 Reference is to Stephen Hawking's statement that because we have laws such as gravity, "the universe can and will create itself from nothing" and there "is no place for a cre-

ator," in various interviews and his book *The Grand Design* (2010).

262   Discovery Channel Special television interview, August 7, 2011 from "Did God Create the Universe?" by Stephen Hawking. Post episode panel discussion.

263   Reference is to Stephen Hawking's statement that because we have laws such as gravity, "the universe can and will create itself from nothing" and there "is no place for a creator," in various interviews and his book *The Grand Design* (2010).

264   Lennox, John. *The Daily Mail.* Accessed 1/24/12 at http:// www.dailymail.co.uk/debate/article-1308599/Stephen-Hawking-wrong-You-explain-universe-God.html   Also see Lennox's book, *God and Stephen Hawking: Whose Design is it Anyway?*

265   Born, Max. (1964) *Natural Philosophy of Cause and Chance.* Mineola, NY: Dover, p.209.

266   Hoyle, Fred (1994) *Home is Where the Wind Blows: Chapters from a Cosmologist's Life,* Mill Valley, CA: University Science Books, p. 421.

267   d'Espagnat, Bernard. (2009) Personal interview at http:// blogs.reuters.com/faithworld/2009/03/17/the-scientist-who-leaves-room-for-spirituality/ accessed 1/19/12.

268   Davies, Paul. "What happened before the big bang," quoted in Stannard, Russell, (2000). *God for the 21st Century*, Philadelphia: Templeton Foundation Press, p. 12.

269   Planck, Max. (Reprinted 1958). *Religion and Naturwissenschaft.* Leipzig: Johann Ambrosius Barth Verlag. XIV Auflage. (Trans. Adam Roman)..

270   Jastrow, Robert. (1992).*God and the Astronomers.* London: WW Norton, Readers Library, p.105.

271   Stoeger, William. "The Origin of the Universe in Science and Religion." From Margenau, & Varghees, eds. (1991). *Cosmos, Bios, Theos: Scientists Reflect on Science, God, and the*

Origins of the Universe, Life, and Homo Sapiens. Peru, IL: Open Court, p. 257.

272 Eddington, Sir Arthur (1923). *TheNatureofthePhysicalWorld*, reproduced by Nabu Press, (2011), Charleston, SC: Nabu, p. 282.

273 Haisch, Bernard. (2010). *The Purpose-Guided Universe: Believing in Einstein, Darwin, and God.* Franklin Lakes, NJ: Career Press, p. 203.

274 Ellis, George. See personal interview with Robert Lawrence Kuhn at http://www.closertotruth.com/video-profile/What-Things-Really-Exist-George-Ellis-/925 accessed 3/12/12

275 Calaprice, & Einstein. (2000). *Expanded Quotable Einstein.* Princeton, NJ: Princeton University Press, p. 207.

276 Russell, Robert John, from "Bridging Science and Religion, Why it Must Be Done." From the Center for Theology and Natural Sciences website, accessed 3/18/12 at http://www.ctns.org/about_history.html#bridging

277 Kant, Immanuel. (1978) *Lectures on Philosophical Theology*, translated by Allen Wood and Gertrude Clark. New York: Cornell University, p. 26.

INDEX

www.ingramcontent.com/pod-product-compliance
Lightning Source LLC
Chambersburg PA
CBHW051317170526
45166CB00002B/574

* 9 7 8 1 4 7 0 0 4 1 5 0 2 *